REVISE BTEC NATIONAL
Engineering

REVISION WORKBOOK

Series Consultant: Harry Smith

Authors: Andrew Buckenham, Kevin Medcalf and Neil Wooliscroft

A note from the publisher

While the publishers have made every attempt to ensure that advice on the qualification and its assessment is accurate, the official specification and associated assessment guidance materials are the only authoritative source of information and should always be referred to for definitive guidance.

This qualification is reviewed on a regular basis and may be updated in the future. Any such updates that affect the content of this Revision Workbook will be outlined at www.pearsonfe.co.uk/BTECchanges.

For the full range of Pearson revision titles across KS2, KS3, GCSE, Functional Skills, AS/A Level and BTEC visit:
www.pearsonschools.co.uk/revise

 Pearson

D1382482

Published by Pearson Education Limited, 80 Strand, London, WC2R 0RL.

www.pearsonschoolsandfecolleges.co.uk

Copies of official specifications for all Pearson qualifications may be found on the website: qualifications.pearson.com

Text and illustrations © Pearson Education Limited 2017
Typeset and illustrated by Kamae Design, Oxford
Produced by Out of House Publishing
Cover illustration by Miriam Sturdee

The rights of Andrew Buckenham, Kevin Medcalf and Neil Wooliscroft to be identified as authors of this work has been asserted by them in accordance with the Copyright, Designs and Patents Act 1988.

First published 2017

20 19 18 17
10 9 8 7 6 5 4 3 2 1

British Library Cataloguing in Publication Data
A catalogue record for this book is available from the British Library

ISBN 978 1 292 15027 7

Printed in Slovakia by Neografia

Acknowledgements
The author and publisher would like to thank the following individuals and organisations for permission to reproduce photographs:

Page 110, "PICAXE VSM software" republished with permission of Revolution Education Ltd.; pages 107, 114, 115, 123, 127, 130, 136, 137, 138, 139 republished with permission of Matrix Technology Solutions Ltd; page 146,Microsoft Excel 2016, Microsoft Corporation

1. While the publishers have made every attempt to ensure that advice on the qualification and its assessment is accurate, the official specification and associated assessment guidance materials are the only authoritative source of information and should always be referred to for definitive guidance.

Pearson examiners have not contributed to any sections in this resource relevant to examination papers for which they have responsibility.

2. Pearson has robust editorial processes, including answer and fact checks, to ensure the accuracy of the content in this publication, and every effort is made to ensure this publication is free of errors. We are, however, only human, and occasionally errors do occur. Pearson is not liable for any misunderstandings that arise as a result of errors in this publication, but it is our priority to ensure that the content is accurate. If you spot an error, please do contact us at resourcescorrections@pearson.com so we can make sure it is corrected.

Websites
Pearson Education Limited is not responsible for the content of any external internet sites. It is essential for tutors to preview each website before using it in class so as to ensure that the URL is still accurate, relevant and appropriate. We suggest that tutors bookmark useful websites and consider enabling students to access them through the school/college intranet.

Introduction

Which units should you revise?

This Workbook has been designed to help you revise the skills you may need for the externally assessed units of your course. Remember that you won't necessarily be studying all the units included here – it will depend on the qualification you are taking.

BTEC Level 3 National Qualification	Externally assessed units
For each of: Extended Certificate Foundation Diploma Diploma	1 Engineering Principles 3 Engineering Product Design and Manufacture
Extended Diploma	1 Engineering Principles 3 Engineering Product Design and Manufacture 6 Microcontroller Systems for Engineers

Your Workbook

Each unit in this Workbook contains either one or two sets of revision questions or revision tasks, to help you **revise the skills** you may need in your assessment. The selected content, outcomes, questions and answers used in each unit are provided to help you to revise content and ways of applying your skills. Ask your tutor or check the Pearson website for the most up-to-date **Sample Assessment Material** and **Mark Schemes** to get an indication of the structure of your actual assessment and what this requires of you. The detail of the actual assessment may change so always make sure you are up to date.

This Workbook will often include one or more useful features that explain or break down longer questions or tasks. Remember that these features won't appear in your actual assessment!

> Grey boxes like this contain **hints and tips** about ways that you might complete a task, interpret a brief, understand a concept or structure your responses.

 This icon will appear next to an example partial answer to a revision question or revision task. You should read the partial answer carefully, then complete it in your own words.

> This is **a revision activity**. It will help you understand some of the skills needed to complete the revision task or question.

> 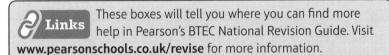 These boxes will tell you where you can find more help in Pearson's BTEC National Revision Guide. Visit **www.pearsonschools.co.uk/revise** for more information.

There is often space on the pages of this Workbook for you to write in. However, if you are carrying out research and make ongoing notes, you may want to use separate paper. Similarly, some units will be assessed through submission of digital files, or on screen, rather than on paper. Ask your tutor or check the Pearson website for the most up-to-date details.

Contents

A small bit of small print

Pearson publishes Sample Assessment Material and the specification on its website. That is the official content, and this book should be used in conjunction with it. The revision questions and revision tasks in this book have been written to help you practise what you have learned in your revision. Remember: the real assessment may not look like this.

Unit 1: Engineering Principles

Your exam

Unit 1 will be assessed through an exam, which will be set by Pearson. You will need to use your ability to solve problems that require individual and combined application of mathematical techniques, and electrical, electronic and mechanical principles to solve engineering problems. You will explore and relate to the engineering contexts and data presented as you respond to short- and long-answer problem-solving questions.

Your Revision Workbook

This Workbook is designed to **revise skills** that might be needed in your exam. The selected content, outcomes, questions and answers are provided to help you revise content and ways of applying your skills. Ask your tutor or check the Pearson website for the most up-to-date Sample Assessment Material and Mark Scheme to get an indication of the structure of your actual exam and what this requires of you. The detail of the actual exam may change so always make sure you are up to date. Make sure you check the instructions in relation to:
- what you need to take into the exam, e.g. a ruler, protractor, pencil, scientific calculator that must not be programmable and that meets the requirements stated
- explaining each stage of your solution and showing all your working so your method is clear when answering any question involving mathematical calculations and stating your answer clearly in the appropriate units
- use of a formulae and constants booklet with reminders of mathematical laws, rules and standard formulae, and constants and engineering formulae – you can find this on pages 47–51 of this Workbook.

To support your revision, this Workbook contains revision questions to help you revise the skills that might be needed in your exam. Each revision paper is divided into three parts:

Section A: Applied mathematics

You will need to demonstrate your knowledge of the algebraic and trigonometric mathematical methods covered in the unit. Each question will have an engineering context but will focus on mathematical methods.

Section B: Mechanical and electrical, electronic principles

You will need to combine good working knowledge of mathematical methods and an understanding of the full range of engineering topics covered in the unit. Each question will cover a single topic from the unit content. You will need to be familiar with the application of a range of formulae to solve engineering problems.

Section C: Synoptic questions

You will give a long answer to a question that draws on knowledge from across the unit. The question is structured in multiple parts that link together different topics from across the unit content. You should answer each part of the synoptic question in the same way that you approach the questions in Section B.

Types of questions

There is guidance in this Workbook for the skills involved in answering the following types of questions.

Describe Explain Find Calculate Solve Draw Label Identify State

Links To help you revise skills that might be needed in your Unit 1 exam, this Workbook contains two sets of revision questions starting on pages 2 and 25. The first is guided and models good techniques to help you develop your skills. The second gives you the opportunity to apply the skills you have developed. See the introduction on page iii for more information on features included to help you revise.

Revision test 1

To support your revision, this Workbook contains revision questions to help you revise the skills that might be needed in your exam. The details of the actual exam may change so always make sure you are up to date. Ask your tutor or check the Pearson website for the most up-to-date Sample Assessment Material to get an idea of the structure of your exam and what this requires of you.

SECTION A: Applied mathematics

Answer ALL questions. Write your answers in the spaces provided.

Complete the guided workings below to give an answer for each of the questions in Section A. Be sure to make use of the information booklet containing formulae and constants on pages 47–51.

1 A laser cutter is being used to make two straight cuts in a sheet of acrylic. The position of the cuts can be represented using a pair of linear simultaneous equations: $8y = 16x + 9$ and $-2y = 8x - 12$

Find the coordinates where the cuts intersect.

When answering 'Find' questions you need to determine the solution to a problem given the information provided. This might involve applying the particular technique or mathematical method mentioned in the question.

Guided

Let $8y = 16x + 9$ be equation (1).

Let $-2y = 8x - 12$ be equation (2).

Links See page 5 of the Revision Guide to revise solving linear simultaneous equations.

Multiply equation (2) by 2 to give: ....

.... Let this be equation (3).

Subtract equation (3) from (1) to give: ....

....

Rearrange to make y the subject: ....

....

Solve for y: ....

....

Substitute y into equation (2) and solve for x: ....

....

Check the answer by substituting x and y into (1):

....

....

Answer: The cuts intersect when $x =$ .... and $y =$ ....

Total for Question 1 = 2 marks

2 A train is accelerating along a level track.

The distance, s, travelled by the train as a function of time, t, is given by the equation $s = 7t + 2t^2$.

Calculate the time taken for the train to travel a distance of 15 m.

> When answering 'Calculate' questions, you need to find the number or amount of something using the information provided in the question. This might involve applying a particular technique, mathematical method or formula.

Substitute $s = 15$ into the equation to give:

$15 = 7t + 2t^2$

Rearrange into the general form of a quadratic, making one side equal to zero:

$2t^2 + 7t - 15 = 0$

Rewrite the equation in the form $2t^2 + 10t - 3t - 15 = 0$

Then: $2t(t + \ldots\ldots\ldots) - 3(t + \ldots\ldots\ldots) = 0$

Take $(t + \ldots\ldots\ldots)$ as a common factor to give:

$(t + \ldots\ldots\ldots)(2t - 3) = 0$

Equate each of the brackets to zero and find two possible values of t:

$(2t - \ldots\ldots\ldots) = 0$ so $t = \ldots\ldots\ldots\ldots\ldots\ldots$

$(t + \ldots\ldots\ldots) = 0$ so $t = \ldots\ldots\ldots\ldots\ldots\ldots$

Check values for t by substituting back into original equation:

$\ldots\ldots\ldots\ldots\ldots\ldots$

$\ldots\ldots\ldots\ldots\ldots\ldots$

The negative value of t is not a feasible response in the given scenario.

Answer: The time taken for the train to travel 15 m is $\ldots\ldots\ldots\ldots\ldots\ldots$

> **Links** See pages 7–8 of the Revision Guide to revise solving quadratic equations.

Total for Question 2 = 2 marks

3 An engineering company manufactures storage tanks from sheet metal. A new design for an open-topped cylindrical tank needs to be analysed to determine the amount of material required in its construction.

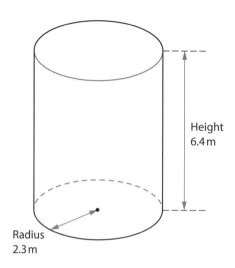

Height
6.4 m

Radius
2.3 m

Calculate the external surface area of the open-topped tank (the material thickness is negligible).

> **Guided**

From the formula booklet, total surface area of a cylinder is: ....

....

But this includes both top and bottom of a fully enclosed cylinder. In this case, only the bottom

is required, so the surface area is: ....

....

Substituting *h* = 6.4 and *r* = 2.3 gives: ....

....

....

Answer: The surface area of the open topped tank A = ...m²

🔗 **Links** See page 14 of the Revision Guide to revise finding the surface area and volume of regular shapes.

Total for Question 3 = 2 marks

4 | The voltage, v_c, across a charging capacitor at time, t, can be represented by the equation:
  | $\ln 12e^{-t} = 2$

Calculate the value of t. Show evidence of the use of the laws of logarithms in your answer.

Guided

Apply the rule $\ln AB = \ln A + \ln B$: ....

....

Apply the rule $\ln A^x = x\ln A$: ....

$\ln 12 - t\ln e = 2$

Since $\ln e = 1$, then: ....

Rearrange to make t the subject: ....

....

....

Solve for t: ....

....

....

Answer: $t =$ ....

Links See page 2 of the Revision Guide to revise applying the rules of logarithms.

Total for Question 4 = 3 marks

5 An engineer has been given a drawing of a triangular steel plate where all the dimensions have not been stated.

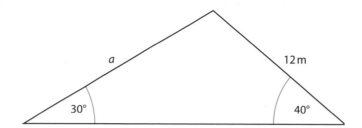

Calculate the length of side *a*.

Guided

From formula booklet, the sine rule is: ....

....

....

Rearrange to make *a* the subject: ....

....

Diagram has been marked up with all sides and angles used in sine rule.

Substitute values into sine rule: ....

....

....

Solve for *a*: ....

....

....

Answer: Length of side *a* is ....

🔗 **Links**   See page 12 of the Revision Guide to revise the application of the sine rule.

Total for Question 5 = 2 marks

END OF SECTION   **TOTAL FOR SECTION A = 11 MARKS**

SECTION B: Mechanical and electrical, electronic principles

Use appropriate units in your answers.

> Complete the guided workings below to give an answer for each of the questions in Section B. Be sure to make use of the information booklet containing formulae and constants on pages 47–51.

6 A sports car with mass 1569 kg accelerates uniformly from rest to a velocity of 100 km/h in a time of 4.7 s.

Calculate the force acting to accelerate the vehicle.

Guided

To find acceleration use $v = u + at$ (from the formula booklet).

Assign values to the variables from information in the question:

$F = ?$

$u = 0$

$v = 100\,km/h =$ .... $= 27.77\,m/s$

$a = ?$

$t = 4.7\,s$

$m = 1569\,kg$

Rearrange to make a the subject: ....

....

Substitute values of v, u and t into formula: ....

$a =$ ....

$a =$ ....m/s^2

To find force use $F = ma$ (from formula booklet).

Substitute values of m and a into the formula: ....

$F =$ ....

$F =$ ....N

Answer: The force acting to accelerate the vehicle is ....N

 Links   See page 22 of the Revision Guide to revise applying the SUVAT equations.

Total for Question 6 = 3 marks

7 A dam is used to hold water in a reservoir. The retaining wall of the dam is 5 m high and 9 m wide. Assume the density of water is 1000 kg/m³.

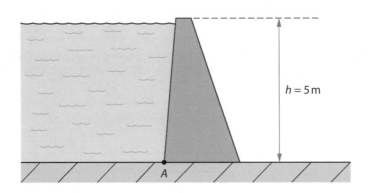

(a) Calculate the hydrostatic thrust acting on the dam.

3 marks

> **Guided**

To find hydrostatic thrust, F, use $F = \rho g A x$ (from formula booklet).

Assign values to the variables from information in the question:

$\rho = 1000$ kg/m³

$g = 9.81$ m/s² (from formula booklet)

$A = 5 \times 9 = 45$ m²

Recall that for a rectangular plane surface the height at which average pressure acts is:

$x = h/2 = 5/2 = 2.5$ m

Substitute values into $F = \rho g A x$:

$F = $ .... × .... × .... × ....

$F = $ .... N

Answer: The hydrostatic thrust acting on the dam is .... N

(b) Calculate the turning moment around the base of the retaining wall.

3 marks

> **Guided**

Replace hydrostatic thrust with an equivalent single point force acting at the centre of pressure. For a rectangular plane surface the centre of pressure is distance $\frac{2}{3}h$ below the free surface

and so distance .... from its base.

To find the moment of a force use $M = Fd$.

Assign values of the variables from the information given in the question:

$F = 1\,103\,625$ N

$d = 5/3 = 1.6$ m

Substitute values into $M = Fd$:

$M = $ .... × .... = ... N m (clockwise)

Answer: The moment around the base of the retaining wall is .... N m

 Links See page 29 of the Revision Guide to revise forces acting on a submerged plane surface.

Total for Question 7 = 6 marks

8 When designing supports for a bridge both vertical and horizontal loading needs to be considered to ensure that the bridge is supported safely.

The diagram shows a simply supported beam in static equilibrium.

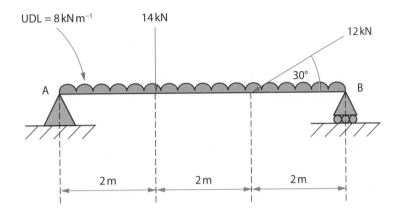

(a) Calculate the vertical and horizontal reaction forces present at the pin support A.

4 marks

> **Guided**

UDL replaced with a single point force acting at the centre of the distribution.

Total force acting along the length of the UDL = $8\,kN/m^1$ x $6\,m$ = $48\,kN$.

12 kN force acting at 30° from the horizontal resolved into vertical and horizontal components

Free body diagram of the beam:

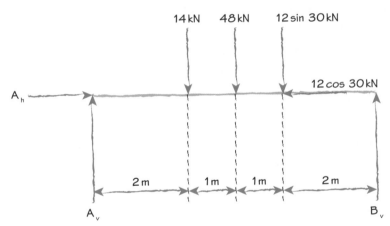

Beam is in static equilibrium so $\sum F_h = 0$ (assume +ve acts left to right).

$\sum F_h = A_h - 12\cos 30 = 0$

$A_h = $ .... kN

Beam is in static equilibrium so $\sum M = 0$.

Find A_v by taking moments about B (assume +ve acting clockwise).

$\sum M_B = (6 \times A_v) - (48 \times 3) - (14 \times 4) - (2 \times 12\sin 30) = 0$

....

....

$A_v = $ .... kN

Answer:

Vertical reaction force at support A is .... kN

Horizontal reaction force at support A is .... kN

(b) Calculate the vertical reaction force present at the roller support B.

> **Guided**

Beam is in static equilibrium so $\sum M = 0$.

Find B_v by taking moments about A (assume +ve acting anticlockwise).

$\sum M_A = (6 \times B_v) - (48 \times 3) - (14 \times 2) - (4 \times 12 \sin 30) = 0$

....

....

$B_v = $ .... kN

Answer: Vertical reaction force at support B is .... kN

Check solution by confirming that $\sum F_v = 0$

$A_v + B_v = 48 + 14 + 12 \sin 30$

.... + ... = 68 kN

> **Links** See page 18 of the Revision Guide to revise simply supported beams.

Total for Question 8 = 6 marks

9 | Waste water flows through a pipe, which increases in diameter from 32 mm to 120 mm.
The initial flow velocity of the water $v_1 = 3$ m/s.

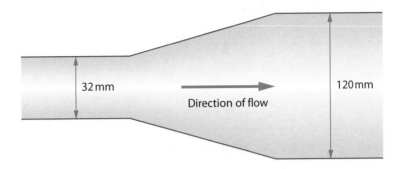

Direction of flow

32 mm   120 mm

Calculate the output flow velocity of the water.

 To find the output flow velocity, v_2, use $A_1 \times v_1 = A_2 \times v_2$ (from the formula booklet).
Rearrange to make v_2 the subject:

$v_2 = $ ....

Assign values to the variables from information in the question:

$A_1 = \dfrac{\pi D_1^2}{4} = $ .... m^2

$A_2 = \dfrac{\pi D_2^2}{4} = $ .... m^2

$v_1 = 3$ m/s

Substitute values into formula: ....

$v_2 = $ ....

....

Answer: The output flow velocity of the waste water is .... m/s

 Links See page 31 of the Revision Guide to revise fluid flow in a gradually tapering pipe.

Total for Question 9 = 4 marks

10 An engineer is analysing a single phase AC waveform on an oscilloscope.
The waveform has periodic time 0.025 s and peak voltage 155.5 V.

(a) Calculate the frequency of the waveform.

1 mark

 To find the frequency of the waveform use $T = \frac{1}{f}$ (from the formula booklet).

$f =$ .... Hz

Answer: The frequency of the waveform is .... Hz

(b) Calculate the root mean square (RMS) voltage.

1 mark

 To find the RMS voltage of the waveform use RMS voltage $= \dfrac{\text{peak voltage}}{\sqrt{2}}$.

RMS voltage $= \dfrac{\text{peak voltage}}{\sqrt{2}}$ (from the formula booklet)

RMS voltage $=$ V

Answer: The RMS voltage of the waveform is V

(c) Explain why the RMS voltage (or current) is useful to engineers working with AC powered systems.

2 marks

When answering 'Explain' questions, you need to give a clear written explanation relating to a particular component, feature or process. You may need to identify a relevant characteristic and then explain its effect.

 The RMS voltage (or current) is equal to an equivalent DC voltage (or current) that when

connected across a .... would produce the same .... effect. This

is useful because RMS values can be used in .... law and power calculations when

dealing with .... loads.

🔗 **Links** See page 72 of the Revision Guide to revise sinusoidal waveform parameters.

Total for Question 10 = 4 marks

11 An ideal transformer used in a USB phone charger has an input voltage of 230 V AC. The transformer output voltage is 5 V AC. The transformer provides a current of 2 A to the charger circuit. There are 200 turns on the secondary transformer coil.

(a) Calculate the number of turns in the primary coil.

2 marks

> **Guided**

To find the number of turns in the primary use $\dfrac{V_1}{V_2} = \dfrac{N_1}{N_2}$ (from the formula booklet).

Assign values to the variables from information in the question:

$V_1 = 230\,V \quad V_2 = 5\,V$

$I_1 = ? \qquad\quad I_2 = 2\,A$

$N_1 = ? \qquad\quad N_2 = 200$

Make N_1 the subject of the formula:

$N_1 = $ ....

Substitute values into the equation:

$N_1 = $ .... = ...

Answer: The primary coil has .... turns.

(b) Calculate the current in the primary coil of the ideal transformer.

2 marks

> **Guided**

In an ideal transformer the power in the primary coil is equal to the power in the secondary coil.

This means that $P = V_1 \times I_1 = V_2 \times I_2$.

Rearrange to make I_1 the subject:

$I_1 = $ ....

Substitute values into the equation:

$I_1 = $ .... = ... A

Answer: The current in the primary coil is .... A

(c) State the two most significant forms of transformer core losses that limit their efficiency.

2 marks

> **Guided**

H .... losses.

E .... currents.

 Links   See page 71 of the Revision Guide to revise mutual inductance and transformers.

Total for Question 11 = 6 marks

12 | An electrical engineer is analysing the network of capacitors shown.

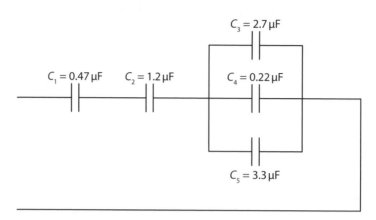

Calculate the total capacitance of the circuit.

Guided

Find the equivalent capacitance of the parallel capacitor network A.

To find total capacitance for parallel network A use $C_A = C_3 + C_4 + C_5$ (from the formula booklet).

C_A = ... μF

Find the equivalent capacitance of the series capacitor network.

To find total capacitance for series network use $\dfrac{1}{C_{Total}} = \dfrac{1}{C_1} + \dfrac{1}{C_2} + \dfrac{1}{C_A}$ (from the formula booklet).

$\dfrac{1}{C_{Total}}$ = ...

C_{Total} = ...

Answer: The total equivalent capacitance of the circuit is .... μF

Links See page 61 of the Revision Guide to revise capacitors in series and parallel combinations.

Total for Question 12 = 2 marks

13

Two AC waveforms can be represented by the equations $v_1 = 25 \sin(30\omega t)$ and $v_2 = 37 \sin\left(30\omega t + \frac{\pi}{3}\right)$.

(a) Draw a labelled sketch of the phasor diagram representing the two waveforms and their resultant.

3 marks

When answering 'Draw' questions, you might need to create a graph or diagram in response to part of a question. This should be done accurately and to an appropriate scale. You may also be asked to label the diagram and provide clear annotations.

(b) Find the equation representing the resultant waveform. Show evidence of the use of the cosine and sine rules in your answer.

5 marks

Guided

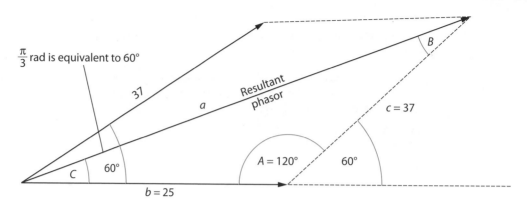

Guided

Find the magnitude of the resultant phasor using the cosine rule $a^2 = b^2 + c^2 - 2bc \cos A$ (from the formula booklet).

Assign values to the variables as identified on the phasor diagram:

$A = 120°$

$b = 25$

$c = 37$

$a = ?$

Substitute values into the formula to find a: ....

$a = $ ....

Find angle of resultant phasor using the sine rule $\dfrac{\sin C}{c} = \dfrac{\sin A}{a}$ (from the formula booklet).

Substitute values into the formula to find C: ....

$C = $ ....

....

Convert value for C from degrees to radians.

$C = $ .... rad

Answer: The resultant waveform is represented by the equation:

$v_R = $ .... $\sin(30\omega t + $ .... $)$

Links See page 74 of the Revision Guide to revise the addition of sinusoidal waveforms.

Total for Question 13 = 8 marks

14 The circuit shows a fully discharged capacitor. When 9 V is applied the capacitor begins to charge through the series resistor.

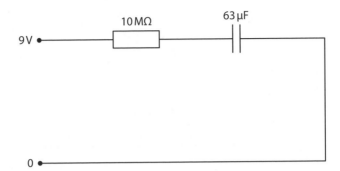

Calculate the voltage across the capacitor 1 minute after the voltage is applied.

Recall that for a charging capacitor $V_c = V_s(1 - e^{\frac{-t}{\tau}})$.

Assign values to the variables from information in the question:

$V_s = 9\,V$

$t = 1\,min = 60\,s$

$\tau = RC = \text{........................} \times \text{........................} = \text{........................}$

Substitute values into the formula:

$V_c = \text{...}$

..

Answer: The voltage across the charging capacitor after 1 min is V

Links See page 54 of the Revision Guide to revise capacitors and RC transients.

Total for Question 14 = 3 marks

15 A simple circuit contains a resistor and diode in series.

The voltage drop across the forward biased silicon diode is 0.7 V.

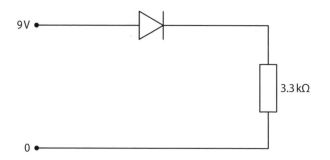

Calculate the current flowing in the circuit.

 To find the voltage across the resistor use Kirchoff's voltage law: $V = V_1 + V_2$ (from the formula booklet).

Assign values to the variables from information in the question:

$V = 9\,V$

$V_1 = 0.7\,V$

$V_2 = ?$

$R = 3.3\,k\Omega = 3300\,\Omega$

$V_2 = \text{.........................} - \text{.........................} = \text{.........................} \, V$

To find the current through the resistor use Ohm's law: $I = \dfrac{V}{R}$ (from the formula booklet).

$I = \text{.........................} = \text{.........................} \, A$

Answer: The current flowing through the resistor is ....

Links See page 60 of the Revision Guide to revise diodes and resistors in series.

Total for Question 15 = 2 marks

16 An engineer specifying an electrical installation must try and minimise resistive losses in long cable runs.

In one application a 38 m run of cable is required to have a maximum resistance of 0.5 Ω.

The resistivity of the copper alloy used in the cable is 1.76×10^{-8} Ω m.

Calculate the conductor cross-sectional area that will meet the maximum resistance requirements.

 Guided

To find the cross-sectional area use $R = \dfrac{\rho l}{A}$ (from the formula booklet).

Assign values to the variables from information in the question:

$l = 38$ m

$R = 0.5$ Ω

$\rho = 1.76 \times 10^{-8}$ Ω m

Rearrange the formula to make A the subject:

A = ...

....

Substitute the values into the equation:

A = ...

....

A = ... m^2

Answer: The cross-sectional area of the conductor is .... m^2

🔗 **Links** See page 43 of the Revision Guide to revise resistance.

Total for Question 16 = 2 marks

17 A power station boiler takes in water at 23 °C.

The water is heated to form super-heated steam at 155 °C, which is then used to run a steam turbine.

The specific heat capacity of water is 4.2 kJ/kg K.

The latent heat of vaporisation for water is 2260 kJ/kg.

The specific heat capacity of dry steam is 1.8 kJ/kg K.

Calculate the heat energy required to heat 1 kg of water from 23 °C to 155 °C.

> **Guided**

Sketch shows a graph of temperature change that occurs with increasing heat energy for water between 23 °C and 155 °C.

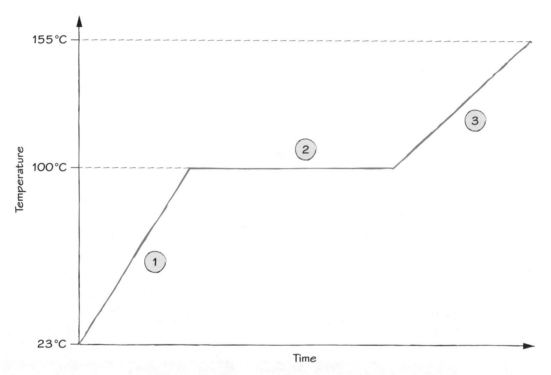

Consider section (1): heating water from 23 °C to boiling point at 100 °C.

To find sensible heat energy use $Q_1 = mc_1\Delta T$.

Assign values to the variables from information in the question:

$m = 1\,kg$

$c_1 = 4.2\,kJ/kg\,K$

$\Delta T = 100 - 23 = $ ....

Substitute the values into the equation:

$Q_1 = $ .... $ = $ .... kJ

Consider section (2): changing phase from liquid to gas.

To find latent heat energy use $Q_2 = ml_2$.

Assign values to the variables from information in the question:

$m = 1\,kg$

$l_2 = 2260\,kJ/kg$

Substitute the values into the equation:

$Q_2 = $ .... $ = $ .... kJ

Consider section (3): heating dry steam from 100 °C to 155 °C.

To find sensible heat energy use $Q_3 = mc_3\Delta T$.

Assign values to the variables from information in the question:

$m = 1\,kg$

$c_3 = 1.8\,kJ/kg\,K$

$\Delta T = 155 - 100 = $ ....

Substitute the values into the equation:

$Q_3 = $ .... $= $ .... kJ

Total energy $Q_{Total} = $ .... $+$ .... $= $ .... kJ

Answer: The total energy required is .... kJ.

 Links See page 35 of the Revision Guide to revise specific heat capacity, latent and sensible heat.

Total for Question 17 = 7 marks

18 A car tyre with fixed volume is inflated to a pressure of 190 kPa at a temperature of 18 °C. After being driven for several hours the temperature of the tyre has increased to 38 °C.

Calculate the pressure in the tyre at the increased temperature of 38 °C.

Apply the general gas equation where:

$$\frac{P_1 V_1}{T_1} = \frac{P_2 V_2}{T_2}$$

When volume is constant:

$$\frac{P_1}{T_1} = \frac{P_2}{T_2}$$

Rearrange to make P_2 the subject of the equation: ....

....

Assign values to the variables from information in the question:

$P_1 = 190 \times 10^3 \, Pa$

$T_1 = 18\,°C = 291\,K$

$T_2 = 38\,°C = 311\,K$

Substitute the values into the equation:

$P_2 = $ .... $ = $ .... Pa

Answer: The pressure in the tyre would be .... kPa

🔗 **Links** See page 40 of the Revision Guide to revise the gas laws.

Total for Question 18 = 2 marks

END OF SECTION **TOTAL FOR SECTION B = 55 MARKS**

SECTION C: Synoptic questions

Answer ALL questions. Write your answers in the spaces provided.

Complete the guided workings below to give an answer for each part of the question in Section C. Be sure to make use of the information booklet containing formulae and constants on pages 47–51.

19 A small portable generator can run for 3.5 hours on a full tank of 2.1 litres of unleaded petrol at its rated output.

The rated output of the generator is 230 V 50 Hz at 3.9 A.

The output of the generator is connected to an external step-down transformer and smoothed full wave rectifier circuit to provide 24 V DC at 34 A.

The equivalent heat energy content of unleaded petrol is 44.4 MJ/kg.

The density of unleaded petrol is 719 kg/m^3.

(a) Calculate the thermal efficiency of the generator.

6 marks

Guided

To calculate efficiency use Efficiency $= \dfrac{\text{power output}}{\text{power input}}$.

Consider energy input in the form of heat energy available in the fuel.

Convert given parameters into standard units:

2.1 L is equivalent to .... m^3

3.5 h is equivalent to 3.5 × 60 × 60 = ... s

Mass of fuel equivalent to 2.1 L in a full tank = 0.0021 m^3 × 719 kg/m^3 = 1.5099 kg

Mass of fuel consumed per second $= \dfrac{1.5099}{12\,600} = $ .... × 10^{-4} kg/s

Equivalent heat energy in petrol consumed per second = ... × 10^{-4} kg/s

        × 44.4 × 10^6 J/kg = ... J/s or W

So, power input = ... W

Consider the energy output in the form of electrical power.

To calculate electrical power use $P = VI$ (from the formula booklet).

Assign values to the variables given in the question:

$I = $ .... A

$V = $ V

Substitute values into the formula:

$P = $ .... × .... = ... W

Efficiency = power output/power Input = ... / .... = 0.16859

Answer: The thermal efficiency of the generator is .... %

(b) Calculate the efficiency of the transformer and rectifier circuit.

3 marks

> **Guided**

To calculate efficiency use Efficiency = $\dfrac{\text{power output}}{\text{power input}}$

Consider power input.

Calculated in part (a): Power input = 897 W

Consider power output.

To calculate electrical power use: $P = VI$ (from the formula booklet).

Assign values to the variables given in the question:

I = ... A

V = ... V

Substitute values into the formula:

P = ... × .... = ... W

Efficiency = $\dfrac{\text{power output}}{\text{power input}}$ = ... / .... = 0.9097

Answer: The efficiency of the transformer is .... %

(c) Draw a circuit diagram for a full wave bridge rectifier circuit with a smoothing capacitor.

5 marks

> **Guided**

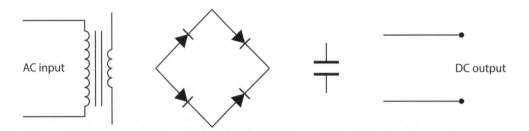

AC input   DC output

Total for Question 19 = 14 marks

END OF SECTION   TOTAL FOR SECTION C = 14 MARKS

END OF PAPER   TOTAL FOR PAPER = 80 MARKS

Revision test 2

To support your revision, this Workbook contains revision questions to help you revise the skills that might be needed in your exam. The details of the actual exam may change so always make sure you are up to date. Ask your tutor or check the Pearson website for the most up-to-date Sample Assessment Material to get an idea of the structure of your exam and what this requires of you.

SECTION A: Applied mathematics

Answer ALL questions. Write your answers in the spaces provided.

Be sure to make use of the information booklet containing formulae and constants on pages 47–51.

1 As part of quality control checks a manufacturer of tension springs carries out experimental testing on samples from each batch manufactured.

   The results for one such test have been plotted as a graph of spring tension, T, versus overall length, l, as shown.

Graph of spring tension versus overall spring length

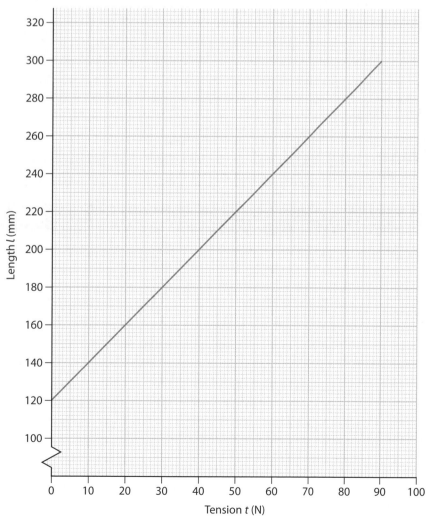

Find the linear equation that represents the relationship between the spring tension, T, and overall length, l.

Answer ....

Links See page 4 of the Revision Guide to revise the equations of lines.

Total for Question 1 = 2 marks

2 A projectile is fired vertically upwards.

The vertical height, h, of the projectile as a function of time, t, is given by the equation $h = t^2 - 6t + 11$.

Find, using factorisation, the values of t when $h = 3$.

Answer..

Links See pages 7–8 of the Revision Guide to revise quadratic equations.

Total for Question 2 = 2 marks

3 A cylindrical oil storage tank has diameter 2.2 m.
The tank is filled to a depth of 0.7 m.

Calculate the volume of oil contained in the tank.

> When answering 'Calculate' questions you need to find the number or amount of something using the information provided in the question. This might involve applying a particular technique, mathematical method or formula.

Answer..

Links See page 14 of the Revision Guide to revise surface area and volumes.

Total for Question 3 = 2 marks

4 The time constant, τ, of a discharging capacitor can be represented by the equation:

$2.5 = 12\,e^{-\tau}$

Solve the equation to find the time constant, τ. Show evidence of the use of the laws of logarithms in your answer.

Remember to use natural logarithms here and that $\ln e = 1$.

When answering 'Solve' questions, you need to determine the solution to a given equation. This might involve applying the particular technique or mathematical method mentioned in the question.

Answer ....

Links See page 2 of the Revision Guide to revise applying the rules of logarithms.

Total for Question 4 = 3 marks

5 A circular steel plate has circumference 2.2 m.
A sector of the circular plate with a subtended angle of 48° is removed by a plasma cutter.

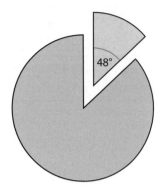

> Remember that angles must be in radians when using the standard formulae for arc length and the area of a sector.

Calculate the remaining area of the steel plate.

Answer ....

 Links See page 9 of the Revision Guide to revise radians, arcs and sectors.

Total for Question 5 = 2 marks

END OF SECTION **TOTAL FOR SECTION A = 11 MARKS**

SECTION B: Mechanical and electrical, electronic principles

Use appropriate units in your answers.

Be sure to make use of the information booklet containing formulae and constants on pages 47–51.

6 The diagram represents the forces acting on the corners of a triangular steel plate.

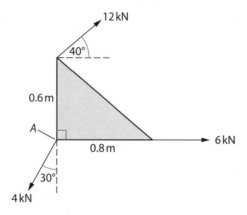

(a) Calculate the magnitude of the resultant force for the system of coplanar non-concurrent forces.

4 marks

Answer ....

(b) Calculate the direction of the resultant force (from the horizontal).

3 marks

Answer ....

(c) Calculate the perpendicular distance from corner *A* to the line of action of the resultant force.

3 marks

Answer ....

Links See pages 15 and 17 of the Revision Guide to revise systems of forces and moments.

Total for Question 6 = 10 marks

7 The diagram shows a steel pin joining two loaded structural members.
The pin has a diameter 25.4 mm.

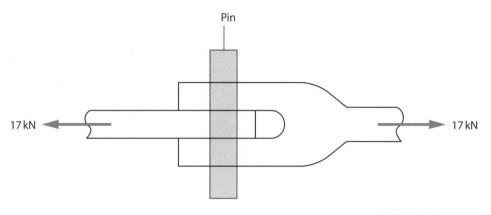

Calculate the shear stress in the pin.

This pin is in 'double shear'.

Answer ....

Links See page 20 of the Revision Guide to revise shear loading.

Total for Question 7 = 3 marks

8 A fence post driver of mass 18 kg is used to hammer in fence posts on soft ground.

The driver is raised a preset distance. It is then released and reaches a velocity of 3.7 m/s as it strikes the post. When the driver strikes the post there is no rebound and the two then move together.

The post itself has a mass of 12 kg.

Calculate the velocity of the post immediately after being stuck.

You will need to apply the principles of conservation of momentum.

Answer ....

Links See page 26 of the Revision Guide to revise Newton's laws.

Total for Question 8 = 3 marks

9 The diagram shows a load being moved vertically up an inclined plane.

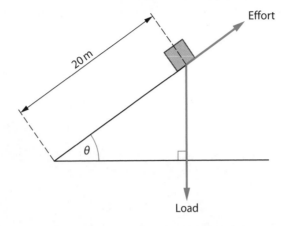

Calculate the value of the angle required for this simple machine to have a velocity ratio of 5.

What is the vertical height moved by the load when the effort has moved 20 m up the slope?

Answer...

Links See page 28 of the Revision Guide to revise mechanical power transmission.

Total for Question 9 = 3 marks

10 | An oil storage tank is 2.6 m wide. It is divided into two compartments by a vertical partition wall across its width. One compartment stores oil with density 840 kg/m³ and depth 0.81 m. The second compartment stores oil with density 910 kg/m³ and depth 1.76 m.

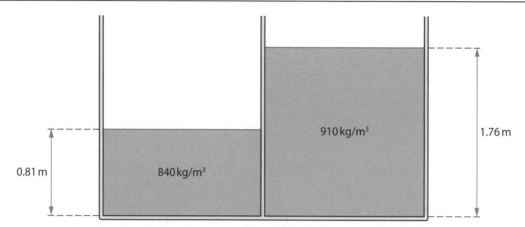

Calculate the resultant hydrostatic thrust on the partition.

Answer...

 Links See page 29 of the Revision Guide to revise submerged surfaces.

Total for Question 10 = 5 marks

11 | An electrical engineer is analysing the network of resistors shown in the circuit diagram.

The circuit shown has a supply voltage of 15 V.

The circuit has total power dissipation of 87.73 mW.

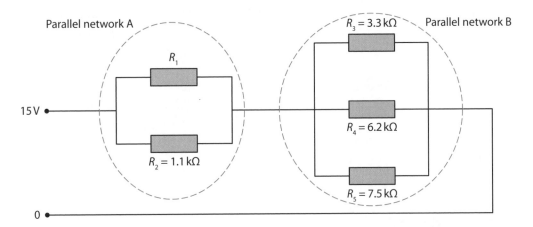

Calculate the value of resistor R_1.

> Start by finding the total network resistance using the power given in the question. Then consider the parallel networks A and B separately.

Answer ....

🔗 **Links**   See page 59 of the Revision Guide to revise resistors in series and parallel combinations.

Total for Question 11 = 6 marks

12 | An accurately machined cylindrical brass bush has a diameter of 215.00 mm at its working temperature of 88 °C.

The coefficient of linear thermal expansion for brass is 18.7×10^{-6} 1/K.

Calculate the diameter of the brass bush when the temperature is reduced to −36 °C.

Answer..

🔗 **Links** See page 34 of the Revision Guide to revise linear expansivity.

Total for Question 12 = 3 marks

13 A rectangular block of high density polyethylene (HDPE) floats in fresh water with density $1000 \, kg/m^3$.

The block is 120 mm long, 80 mm wide and 50 mm thick.

The block is partially submerged and 47.8 mm of its height is below the free surface of the water.

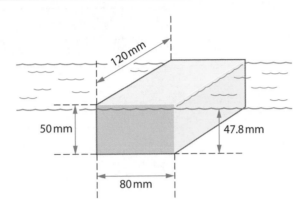

Calculate the density of the HDPE.

Answer ....

Links See page 30 of the Revision Guide to revise immersed bodies.

Total for Question 13 = 3 marks

14 | A battery generates an emf of 9.29 V.
When connected to a resistive load of 370 Ω, a current of 25 mA flows around the circuit.

(a) Calculate the internal resistance of the battery.

2 marks

It might help to sketch a quick circuit diagram showing the internal resistance and load as separate resistors fed by a 9.29 V supply.

Answer..

(b) Calculate the voltage present at the battery terminals.

2 marks

Answer..

Links   See page 57 of the Revision Guide to revise DC power sources.

Total for Question 14 = 4 marks

15 | Magnetic screening is used to isolate sensitive electronic devices from the effects of magnetic fields.

State one important characteristic essential for materials to be effective as magnetic screens.

....

 Links See page 66 of the Revision Guide to revise reluctance and magnetic screening.

Total for Question 15 = 1 mark

16 | Electric motors are in widespread use in thousands of engineering applications. They are used to convert electrical power into mechanical rotation.

Explain the function of the commutator in a DC electric motor.

> When answering 'Explain' questions, give a clear written explanation relating to a particular component, feature or process. You may need to identify a relevant characteristic and then explain its effect.

....

....

....

....

....

....

Links See page 68 of the Revision Guide to revise DC motors.

Total for Question 16 = 4 marks

17 | An emf is induced in a coil when it rotates inside a magnetic field. This is the basic principle by which electric generators operate.

State four factors that effect the magnitude of the emf generated by a coil rotating in a magnetic field.

1 ....

2 ....

3 ....

4 ....

Links See page 69 of the Revision Guide to revise electric generators.

Total for Question 17 = 4 marks

18 | A solenoid is 120 mm long and is wound with 800 turns.

Calculate the current required for a magnetic field strength of 36 000 A/m to be generated inside the solenoid.

Answer..

Links See page 63 of the Revision Guide to revise magnetic fields.

Total for Question 18 = 2 marks

19 | An engineer is analysing an AC circuit containing both resistance and capacitance.

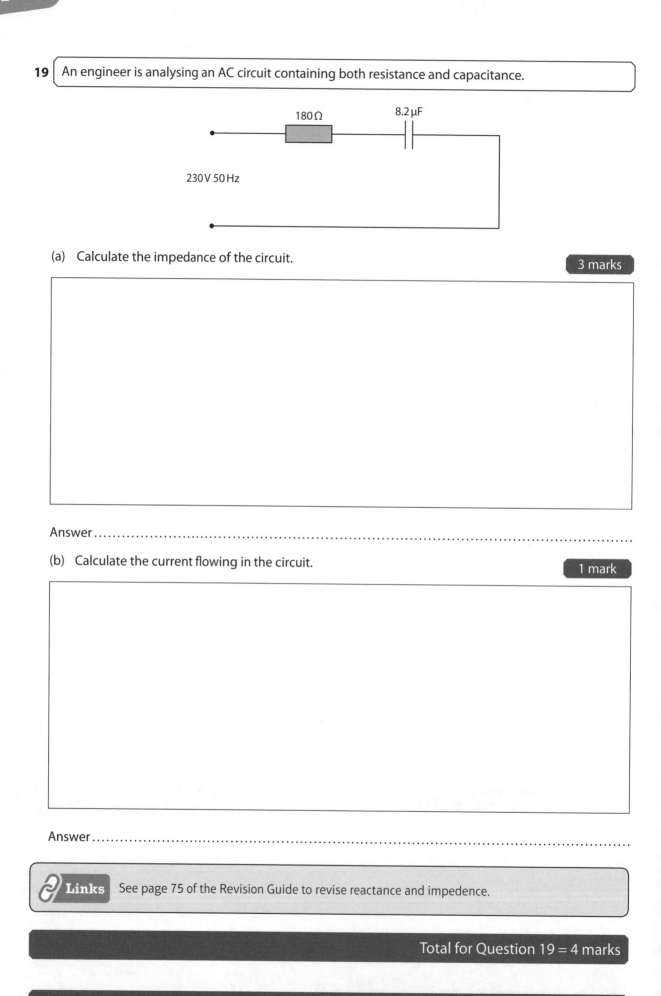

180 Ω   8.2 µF

230 V 50 Hz

(a) Calculate the impedance of the circuit.

3 marks

Answer ....

(b) Calculate the current flowing in the circuit.

1 mark

Answer ....

Links See page 75 of the Revision Guide to revise reactance and impedence.

Total for Question 19 = 4 marks

END OF SECTION TOTAL FOR SECTION B = 55 MARKS

SECTION C: Synoptic questions

Answer ALL questions. Write your answers in the spaces provided.

Be sure to make use of the information booklet containing formulae and constants on pages 47–51.

20 A vehicle undergoing testing is driven uphill from a standing start along a straight track.

The vehicle is powered by an internal combustion engine, which runs on petrol.

During the test:

- the vehicle accelerates at an average $4\,m/s^2$ over a distance of $160\,m$
- the vehicle gains $20\,m$ in height
- the vehicle's $12\,V$ electrical systems require an average current of $16.58\,A$
- petrol is consumed at an average rate of $47.2\,g/s$.

The car has a mass of $1860\,kg$.

Equivalent heat energy content for petrol is $42.4\,MJ/kg$.

(a) Find the total useful work done by the internal combustion engine during the test. [10 marks]

Useful work will include increases in kinetic and potential energy and the electrical energy consumed during the test.

When answering 'Find' questions, you need to determine the solution to a problem given the information provided in the question. This might involve applying the particular technique or mathematical method mentioned in the question.

Answer ....

(b)   Calculate the overall efficiency of the vehicle during the test.

4 marks

Answer...

Total for Question 20 = 14 marks

END OF SECTION   **TOTAL FOR SECTION C = 14 MARKS**

END OF PAPER   **TOTAL FOR PAPER = 80 MARKS**

Formulae and constants

During your exam for Unit 1 you will have access to a booklet containing a list of formulae and physical constants. You must become familiar with this booklet, know what is included and what you will need to remember. This booklet is shown on pages 47-51 and is included in the Sample Assessment Material for Unit 1 on the BTEC Nationals Engineering section of the Pearson website. Always check the website to ensure you are up to date.

Formulae and Constant

Static and Direct Current electricity theory

Current $\qquad I = \frac{q}{t}$

Coulomb's law $\qquad F = \frac{q_1 q_2}{4\pi\varepsilon_0 r^2}$

Resistance $\qquad R = \frac{\rho l}{A}$

Resistance: temperature coefficient $\qquad \frac{\Delta R}{R_0} = \alpha\Delta T$

Ohm's Law $\qquad I = \frac{V}{R}$

Total for resistors in series $\qquad R_T = R_1 + R_2 + R_3$

Total for resistors in parallel $\qquad \frac{1}{R_T} = \frac{1}{R_1} + \frac{1}{R_2} + \frac{1}{R_3} \dots$

Power $\qquad P = IV, \quad P = I^2 R, \quad P = \frac{V^2}{R}$

Efficiency $\qquad E = \frac{P_{out}}{P_{in}}$

Kirchhoff's current law $\qquad I = I_1 + I_2 + I_3$

Kirchhoff's voltage law $\qquad V = V_1 + V_2 + V_3 \ \text{ or } \ \sum PD = \sum IR$

Capacitance

Electric field strength $\qquad E = \frac{F}{q}$

Electric field strength: uniform electric fields $E = \frac{V}{d}$

Capacitance $\qquad C = \frac{\varepsilon A}{d}$

Time constant $\qquad \tau = RC$

Charged stored $\qquad Q = CV$

Energy stored in a capacitor $\qquad W = \frac{1}{2}CV^2$

Capacitors in series $\qquad \frac{1}{C_T} = \frac{1}{C_1} + \frac{1}{C_2} + \frac{1}{C_3} \dots$

Capacitors in parallel $\qquad C_T = C_1 + C_2 + C_3$

Voltage decay on capacitor discharge $\qquad v_c = Ve^{(-t/\tau)}$

Magnetism and electromagnetism

Magnetic flux density $\qquad B = \frac{\phi}{A}$

Magneto motive force $\qquad F_m = NI$

Magnetic field strength or magnetising force $H = \frac{NI}{l}$

Permeability	$\frac{B}{H} = \mu_0\mu_r$
Reluctance	$S = \frac{F_m}{\phi}$
Induced EMF	$E = BLv, \; E = -N\frac{d\phi}{dt} = -L\frac{dI}{dt}$
Energy stored in an inductor	$W = \frac{1}{2}LI^2$
Inductance of a coil	$L = \frac{N\phi}{I}$
Transformer equation	$\frac{V_1}{V_2} = \frac{N_1}{N_2}$

Single phase Alternating Current theory

Time period	$T = \frac{1}{f}$
Capacitive reactance	$X_C = \frac{1}{2\pi fC}$
Inductive reactance	$X_L = 2\pi fL$
Root mean square voltage	$RMS\ voltage = \frac{peak\ voltage}{\sqrt{2}}$

Total impedance of an inductor in series with a resistance	$Z = \sqrt{X_L^2 + R^2}$

Total impedance of a capacitor in series with a resistance	$Z = \sqrt{X_C^2 + R^2}$

Average waveform value average value	$Average\ value = \frac{2}{\pi} \times maximum\ value$
Form factor of a waveform	$Form\ factor = \frac{RMS\ value}{average\ value}$

Laws of Mathematics

Rules of indices

$a^m \times a^n = a^{(m+n)}$

$a^m \div a^n = a^{(m-n)}$

$(a^m)^n = a^{mn}$

Rules of logarithms

$\log AB = \log A + \log B$

$\log\frac{A}{B} = \log A - \log B$

$\log A^x = x\log A$

Trigonometric rules

Sine rule

$$\frac{a}{\sin A} = \frac{b}{\sin B} = \frac{c}{\sin C} \text{ or } \frac{\sin A}{a} = \frac{\sin B}{b} = \frac{\sin C}{c}$$

Cosine rule

$$a^2 = b^2 + c^2 - 2bc \cos A$$

Volume and area of regular shapes

length of an arc of a circle	$s = r\theta$
area of a sector of a circle	$A = \frac{1}{2}r^2\theta$
volume of a cylinder	$V = \pi r^2 h$
total surface area of a cylinder	$TSA = 2\pi rh + 2\pi r^2$
volume of sphere	$V = \frac{4}{3}\pi r^3$
surface area of a sphere	$SA = 4\pi r^2$
volume of a cone	$V = \frac{1}{3}\pi r^2 h$
curved surface area of cone	$CSA = \pi rl$

Quadratic formula

To solve $ax^2 + bx + c = 0$, $a \neq 0$

$$x = \frac{-b \pm \sqrt{b^2 - 4ac}}{2a}$$

Equations of linear motion with uniform acceleration

$$v = u + at$$

$$s = ut + \frac{1}{2}at^2$$

$$v^2 = u^2 + 2as$$

$$s = \frac{1}{2}(u + v)t$$

Stress and strain

Direct stress	$\sigma = \dfrac{F}{A}$
Direct strain	$\varepsilon = \dfrac{\Delta L}{L}$
Shear stress	$\tau = \dfrac{F}{A}$
Shear strain	$\gamma = \dfrac{a}{b}$
Modulus of elasticity	$E = \dfrac{\sigma}{\varepsilon}$
Modulus of rigidity	$G = \dfrac{\tau}{\gamma}$

Work, power, energy and forces

Force	$F = ma$
Resultant force	$F_x = F\cos\theta,\ F_y = F\sin\theta$
	(where θ is measured from the horizontal)
Mechanical work	$W = Fs$
Force to overcome limiting friction	$F = \mu N$
Gravitational potential energy	$PE = mgh$
Kinetic energy	$KE = \dfrac{1}{2}mv^2$

Gas laws

Boyle's Law	$pV = \text{constant}$
Charles's Law	$\dfrac{V}{T} = \text{constant}$
General gas equation	$\dfrac{pV}{T} = \text{constant}$

Angular parameters

Centripetal acceleration	$a = \omega^2 r = \dfrac{v^2}{r}$

Power $\qquad P = T\omega$

Rotational Kinetic energy $\quad KE = \frac{1}{2}I\omega^2$

Angular frequency $\qquad \omega = 2\pi f$

Frequency $\qquad f = \dfrac{1}{\text{time period}}$

2π radians $= 360°$

Physical constants

Acceleration due to gravity $\qquad g = 9.81\text{m/s}^2$

Permittivity of free space $\qquad \varepsilon_0 = 8.85 \times 10^{-12}\,\text{F/m}$

Permeability of free space $\qquad \mu_0 = 4\pi \times 10^{-7}\,\text{H/m}$

Thermodynamic principles

Sensible heat $\qquad Q = mc\Delta T$

Latent heat $\qquad Q = ml$

Entropy and enthalpy $\quad H = U + pV$

Linear expansivity $\qquad \Delta L = \alpha L\Delta T$

Fluid principles

Continuity of volumetric flow $\qquad A_1v_1 = A_2v_2$

Continuity of mass flow $\qquad \rho A_1v_1 = \rho A_2v_2$

Hydrostatic thrust on an immersed plane surface $\quad F = \rho gAx$

Unit 3: Engineering Product Design and Manufacture

Your set task

Unit 3 will be assessed through a task, which will be set by Pearson. You will need to use your understanding of engineering product design and manufacturing processes, considering function, sustainability, materials, form and other factors. You will complete a task that requires you to follow a standard development process of interpreting a brief, carrying out research, scoping initial design ideas, preparing a design proposal and evaluating your proposal.

Your Revision Workbook

> This Workbook is designed to **revise skills** that might be needed in your assessed task. The selected content, outcomes, questions and answers are provided to help you revise content and ways of applying your skills. Ask your tutor or check the Pearson website for the most up-to-date Sample Assessment Material and Mark Scheme to get an indication of the structure of your actual assessed task and what this requires of you. The detail of the actual assessed task may change so always make sure you are up to date. Make sure you check the instructions in relation to having a pen, pencil, ruler, eraser, drawing instruments and calculator.

To support your revision, this Workbook contains a revision task to help you revise the skills that might be needed in your assessed task. The revision task is divided into sections.

Researching and making notes

You will use your skills to read a task brief and research into engineering design and manufacture in relation to a product, and make notes.

Reviewing further information

You will then interpret further information in relation to the product and the design and manufacturing processes to be considered.

Responding to activities

Your response to the brief will involve you in the following activities:
- project planning and product design changes made during an iterative development process
- interpreting a brief into operational requirements
- producing a range of initial design ideas based on the client brief
- developing a modified product proposal with relevant design documentation
- validating the design proposal.

> To help you revise for your Unit 3 set task, this Workbook contains a **full revision task**. See the introduction on page iii for more information on features included to help you revise.

Revision Task

To support your revision, this Workbook contains a full revision task to help you revise the skills that might be needed in your assessed task. The details of the actual assessed task may change so always make sure you are up to date. Ask your tutor or check the Pearson website for the most up-to-date Sample Assessment Material to get an idea of the structure of your assessed task and what this requires of you. Start by reading the revision task brief below carefully.

Task brief

A manufacturer has been approached by one of its clients for whom it manufactures a jointing system for wooden roof beams. The system consists of two steel plates and a fixing kit.

The client has asked the manufacturer to optimise the design of the jointing system. The optimised solution should minimise the number of separate components required and be quick and easy to install.

You will research and prepare notes on the possible design, materials and manufacturing processes to achieve this.

 Now read the task information below carefully. Then read the brief and information again, underlining key information.

Task information

The jointing system is used to fix wooden beams end to end by sandwiching the joint between the two metal plates.

The two low carbon steel plates are attached to either side of the joint using 6 off M12 medium carbon steel bolts and associated nuts, plain washers and spring washers. Fitting involves drilling six holes in the end of the beams, three on each side of the joint.

The metal plates are manufactured in batches of two thousand at a time.

Individual metal plates are laser cut from a large mild steel plate.

The plates are galvanised at the end of the manufacturing process.

The jointing system must be suitable for use in a wide range of environmental conditions and temperatures ranging from −40 °C to +50 °C.

Plate dimensions: 300 × 95 × 5. All dimensions in mm.

Researching and making notes

Planning your time

When carrying out research you need to break it into stages. Estimate how long you will need for each stage, then plan and monitor your time to ensure you can complete everything within the time allocated. The stages involved in research in this Workbook are noted below. For your actual assessment, check the Sample Assessment Material on the Engineering page of the Pearson website for details in relation to research, notes, and timing.

Breakdown of task	Time
Description and analysis of existing design and identify important areas for research	
Research possible product designs, including each important area	
Research suitable materials, processes, fixtures, fittings, relevant numerical data	

Identifying areas for research

Guided

Complete the following idea map by:
- writing a short description of the **existing product design**
- identifying possible **areas for research.** Your research focus will reflect the brief and task information and might include: how the existing product could be improved; optimising the existing design; similar existing products, manufacturing methods and processes, costs, materials and material properties; health and safety issues; sustainability and environmental issues; relevant numerical data; aesthetics; ergonomics; anthropometrics; joints/fixings; applied finish; and advantages/disadvantages of the design.

You may wish to summarise information in the diagram and use additional paper for more detailed notes.

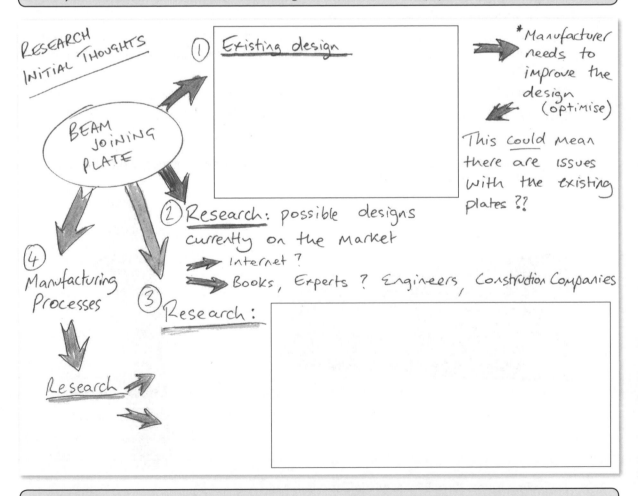

To revise interpreting a brief and planning research, see pages 154–155 of the Revision Guide.

Product research and notes

 Use the internet to find at least **three different current products** that are available for **joining wooden roof beams**. The extract below is an example of notes on one product, to help guide your further product research. Make sure you include all the important areas in your research.

Product 1 – example

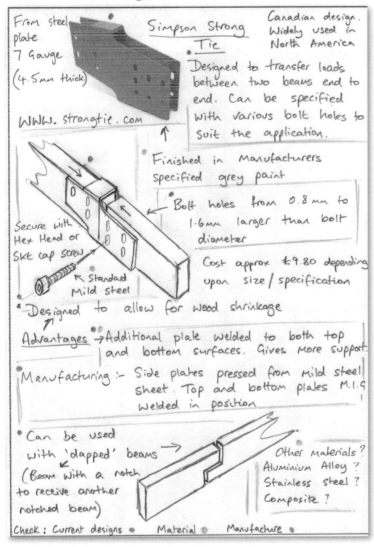

From steel plate 7 Gauge (4.5mm thick)

WWW. strongtie.com

Secure with Hex Head or Skt cap screw

↖ Standard Mild steel

* Designed to allow for wood shrinkage

Advantages → Additional plate welded to both top and bottom surfaces. Gives more support.

Manufacturing :- Side plates pressed from mild steel sheet. Top and bottom plates M.I.G welded in position

* Can be used with 'dapped' beams → (Beam with a notch to receive another notched beam)

Simpson Strong Tie

Canadian design. Widely used in North America

* Designed to transfer loads between two beams end to end. Can be specified with various bolt holes to suit the application.

Finished in Manufacturers specified grey paint

* Bolt holes from 0.8mm to 1.6mm larger than bolt diameter

Cost approx £9.80 depending upon size / specification

Other materials? Aluminium Alloy? Stainless steel? Composite?

Check: Current designs • Material • Manufacture •

Focus for further product research

When finding products for research, look for designs that use methods similar to the ones used by the manufacturer. You could search for 'joist repair', 'roof beam join plates' or 'joining joists end to end'. The diagram below shows some alternative methods you may find from an internet search. Pay particular attention to those that are simple to fit, don't require to be held in place during fitting and use as few different component parts as possible.

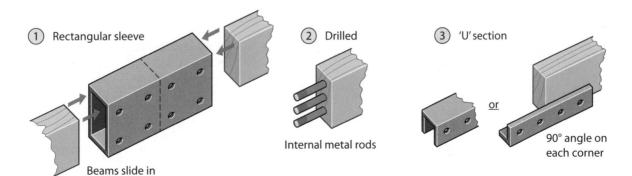

① Rectangular sleeve

Beams slide in

② Drilled

Internal metal rods

③ 'U' section

or

90° angle on each corner

Product 2

  Guided

 Use this page to make research notes on your chosen product 2. Consider how the existing product could be improved and aspects of the existing design that could be optimised.

Links To revise research notes, see page 156 of the Revision Guide.

Images of possible solution and function of product:

Manufacturing methods and processes:

Costs:

Materials and material properties:

Health and safety issues:

Sustainability and environmental factors:

Relevant numerical data:

Aesthetics, ergonomics, anthropometrics:

Applied finish:

Advantages/disadvantages of the design:

Product 3

Use this page to make research notes on product 3. You can use the example on page 55 and the headings on page 56 to help structure your research.

My research notes

Product 4

Use this page if you want to make research notes on product 4. You can use the example on page 55 and the headings on page 56 to help structure your research.

My research notes

Researching materials, processes, fixtures, fittings and relevant numerical data

> You need to carry out any research that is specifically required by the brief you have been given.
>
> **Use the next two pages to make notes on:**
>
> 1. roof beam materials and beam dimensions; 2. nail plates; 3. materials; 4. fasteners; 5. welding processes.

> Guided >

> ✎ Carry out research on common **roof beam materials** and **beam dimensions**. Try looking for roof beams, joists or rafters. Make notes on the mechanical properties of the materials used and the range of standard sizes commonly used in construction.

The most popular wood species for roof joists and rafters is ....

....

....

In the UK construction industry, beam dimensions are strictly specified according to the load they

are expected to support. It also depends upon the grade of timber used. For example, ...

....

....

....

....

> ✎ Carry out research on **nail plates.** Make notes on how they work, the materials used, how they are manufactured, how easy they are to install and their overall strengths and weaknesses.

Nail plates are the most popular joist joining method in the UK. Most are manufactured in galvanised

1mm steel plate. ....

....

....

> ✎ Research **low carbon steel** and potential alternative materials like **aluminium alloy duralumin** and **austenitic stainless steel alloy**. Make notes on their mechanical properties and other factors that might be important, such as corrosion resistance, cost and the effects of low temperatures.

The materials used for the joining plates might have to operate in temperatures as low as $-40\,°C$, depending where in the world the building is situated.

My research has shown that low carbon steel can lose some of its impact strength at extremely low temperatures, making it brittle. Steel operating below the ductile to brittle transition temperature is far more likely to fail under shock loading.

However, duralumin and austenitic stainless steel alloys retain their impact strength and resistance to shock loading even at very low temperatures.

....

....

....

....

....

....

....

....

....

....

....

> ✏️ Research a range of **fasteners**, such as **socket cap screws, structural bolts, washers** and **coach bolts.**
> Comment upon speed of assembly, load spreading ability and strength of the different options.

....

....

....

....

....

....

....

....

....

....

....

....

> ✏️ Research the **MIG** and **TIG welding processes** as possible manufacturing methods for a new joist
> joining bracket.

....

....

....

....

....

....

....

....

....

....

Using preparatory notes

In this **Revision Workbook** you can refer to any of the notes you have made as you give answers to the activities that follow. In your **actual assessment**, you may not be allowed to refer to notes, or there may be restrictions on the length and type of notes that are allowed. Check with your tutor or look at the most up-to-date Sample Assessment Material on the Pearson website for information.

Use the checklist below for the research and notes you have completed in this Workbook.

☐ I have conducted research that is directly relevant to the task.

☐ I have considered how the existing product could be improved.

☐ I have anticipated aspects of the existing design that could be optimised.

☐ I have thoroughly researched other similar products on the market.

☐ I have made research notes that include:

- the function of the product
- manufacturing methods and processes
- costs
- materials and material properties
- health and safety issues
- sustainability and environmental issues
- relevant numerical data
- aesthetics
- ergonomics
- anthropometrics
- joints / fixings
- applied finish
- advantages/disadvantages of the design.

Reviewing further information

Review the further information below that repeats the product images from page 53 and provides **additional images and information**. Then read the **further task** information on pages 63–64 that relates to these images.

(a)

(b)

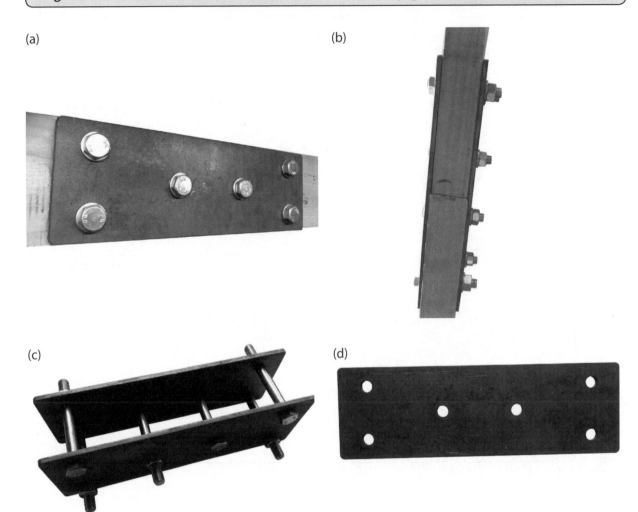

(c)

(d)

Plate dimensions: 300 × 95 × 5. All dimensions in mm.

(e)

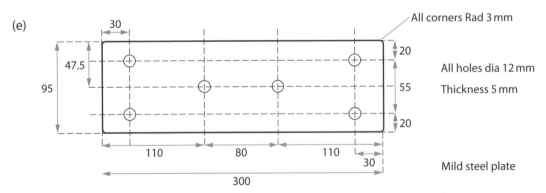

All corners Rad 3 mm

All holes dia 12 mm

Thickness 5 mm

Mild steel plate

Parts list

Name	Description	Quantity
Plates	Low carbon steel joining plates	2
Bolts	M12 × 1.75 × 60 mm	6
Nut	M12	6
Plain washer	M12	6
Spring washer	M12	6

Read the task information below carefully. It repeats the task brief and information from page 53 and provides **further information** with the **client brief**, which relates to the images on page 62. Underline the key information.

Task brief

A manufacturer has been approached by one of its clients for whom it manufactures a jointing system for wooden roof beams. The system consists of two steel plates and a fixing kit.

The client has asked the manufacturer to optimise the design of the jointing system. The optimised solution should minimise the number of separate components required and be quick and easy to install.

You will research and prepare notes on the possible design, materials and manufacturing processes to achieve this.

Task information

The jointing system is used to fix wooden beams end to end by sandwiching the joint between the two metal plates.

The two low carbon steel plates are attached to either side of the joint using 6 off M12 medium carbon steel bolts and associated nuts, plain washers and spring washers. Fitting involves drilling six holes in the end of the beams, three on each side of the joint.

The metal plates are manufactured in batches of two thousand at a time.

Individual metal plates are laser cut from a large mild steel plate.

The plates are galvanised at the end of the manufacturing process.

The jointing system must be suitable for use in a wide range of environmental conditions and temperatures ranging from −40 °C to +50 °C.

Further task information

Client brief

Images (a) and (b) show the jointing system in situ joining two beams together.

Image (c) shows an assembled jointing system that is not fitted to a beam, showing how the fasteners hold the two plates in place.

Image (d) shows a single joining plate (before it has been galvanised).

Image (e) shows an orthographic engineering drawing of a single joining plate.

The client is aware that the current design has a number of issues, but the redesign has been triggered by the plates fracturing in service when used in extreme environmental conditions.

The client had intended the life cycle of the plates to be at least 50 years. The client needs the manufacturer to identify the stage in the life cycle when the plates begin to exhibit signs of fracture and design a solution that will reduce the likelihood of the plates failing in service.

Based on extensive simulations and testing, the client has provided the following information in Table 1, which can be used to perform a statistical analysis of the service conditions in which the plates are used.

The client has asked the manufacturer to come up with an alternative solution that takes into account the most efficient and sustainable use of materials and manufacturing processes. The manufacturer also has an opportunity to optimise the design of the jointing system to make it quicker and easier to fit. This will involve combining the function of the two separate plates into a single component.

The method of joining the wooden beams must:

1. Join two beams (of section 95 × 30 mm) end to end.

2. Not protrude further than 40 mm from the top/bottom/side faces of the beams.

Table 1: Outcome of simulations and testing on existing plates

Plate	Location	Min temp inside building °C	Max temp inside building °C	Average humidity %	Life cycle (years)							
					Test 1	Test 2	Test 3	Test 4	Test 5	Test 6	Test 7	Test 8
A	Brazil	8	44	80	40	48	55	51	49	49	50	56
B	Spain	6	40	78	40	47	52	52	51	44	46	47
C	U.K.	−4	28	50	39	39	41	37	40	34	31	37
D	Norway	−15	24	40	26	29	29	29	27	23	20	28
E	Alaska	−35	24	40	11	20	14	18	16	19	20	20

When you have read the additional information, you could make brief key notes that interpret it.

For example, with this client brief and information:

1. State the average expected life cycle of the product.

2. What is the diameter of the holes in the existing product?

3. Identify the main client motivation for redesigning the product.

1 ....

2 ....

3 ....

 Links To revise data analysis, see pages 145–149, 159 and 163 of the Revision Guide.

When you have understood the task information, work through the five revision activities:

Page 65: Revision activity 1: A record of project planning and design changes made during the iterative development process.

Page 68: Revision activity 2: Interpretation of the brief into operational requirements.

Page 70: Revision activity 3: A range of initial ideas based on the client brief.

Page 74: Revision activity 4: A modified product proposal with relevant design documentation.

Page 79: Revision activity 5: An evaluation of the design proposal.

Responding to activities

To answer these revision activities, you will have carried out research in relation to the revision task brief and information on page 53 and the further information on pages 62–64, which includes a client brief, engineering drawings and data. In this **Revision Workbook** you can refer to any of the notes you have made as you give answers to the activities. In your **actual assessment** you may not be allowed to refer to notes, or there may be restrictions on the length and type of notes that are allowed. Check with your tutor or look at the most up-to-date Sample Assessment Material on the Pearson website for information.

Revision activity 1: Planning and design changes made during the iterative development process

At the start of the task create a short outline project time plan in your Workbook.

During the iterative development process you should also record in your Workbook:

1. Why changes were made to the design during each session.
2. Action points for the next session.

Make sure that:
- your plans and records are **logical** and show an **iterative** approach to the design process
- the design development activities lead to **refinements** that link to **research** and the **requirements** of the brief
- changes made to all design developments are **justified**
- identified action points for the next session are **logical**, **prioritised** and **SMART**: Specific, Measurable, Achievable, Realistic, and Time-based.

 Guided

To **create a short outline project time plan** to show how you intend to use the time available for the activities, find out how much time you are allowed for activities in the actual assessment by asking your tutor or looking at the Sample Assessment Material on the Pearson website. Then, make your own plan below, breaking down the activities and time into detail. Use a format that best works for you, e.g. Gantt chart, timeline, flowchart of written list of tasks.

My project time plan

 Links
To revise time planning, see page 160 of the Revision Guide, and for iterative development see pages 144 and 166 of the Revision Guide.

To **record changes and action points** for the development process, use the pages that follow. For each session, record why changes were made and action for the next session. Action points should show forward planning that is clearly linked to the specifics of the product being redesigned, with consideration of what has happened in the previous session. Explain and justify the specific changes made in order to fulfil the requirements of the client brief. The first entry has been completed as an example.

Session 1: Date: Why design changes were made: As part of turning the brief into a list of operational requirements I analysed the numerical data given in the brief. It looks like use at low temperatures increases the chance of existing plates failing. I used my research on materials to look at ductile/brittle transition temperature for steel and now think the design should be changed to use a different material that won't become brittle in the cold.

Action for next session: I need to look at how I can make the jointing system easier to fit. I will use my research on existing beam joining products to help come up with more ideas.

Session 2: Date: Why design changes were made: ....

....

....

....

Action for next session: ....

....

....

....

Session 3: Date: Why design changes were made: ....

....

....

....

Action for next session: ....

....

....

....

Session 4: Date: Why design changes were made: ....

....

....

....

Action for next session: ....

....

....

....

Session 5: Date: Why design changes were made: ....

....

....

....

Action for next session: ....

....

....

....

Session 6: Date: Why design changes were made: ....

....

....

....

Action for next session: ....

....

....

....

Session 7: Date: Why design changes were made: ....

....

....

....

Revision activity 2: Interpret the brief into operational requirements

Interpret the brief into operational requirements, to include:

- product requirements
- opportunities and constraints
- interpretation of numerical data
- key health and safety, regulatory and sustainability factors.

Make sure that:
- your product requirements are **cohesive** and **comprehensive**
- the opportunities and constraints are **feasible** and **meet the brief**, **enhancing** product performance
- your calculation and interpretation of numerical data is **accurate** and conclusions are commented upon/ taken forward
- the health and safety, regulatory and sustainability factors are **relevant** to the given context with the redesign of the product in mind.

> **Guided**

Read the **further information** and **client brief** carefully (pages 62–64) and make a list of all the **product requirements**. The first one has been done for you.

The product must join wooden beams of section 95 × 30, end to end.

....

....

....

....

Make a list of all the **opportunities** and **constraints**. The first one for each has been done for you. Remember, you can always come back and add to these once you have fully analysed the task.

Opportunities

1 Joining the beams without drilling through them could be a better solution than the existing design.

....

....

....

....

Constraints

1 The chosen solution must not protrude from any face of the beams by more than 40mm.

....

....

....

....

Interpret the data in Table 1 (page 64) to show understanding of the client brief by completing the sentences below.

The mean life cycle in years for plates used in the United Kingdom is ....

The mean life cycle in years for plates used in Norway is ....

The country where beam plates have the shortest average life cycle is ....

The country where beam plates have the longest average life cycle is ....

The temperature range the plates are subjected to is least in ....

The temperature range the plates are subjected to is the biggest in ....

The plates are subjected to the lowest average humidity in ....

The data would seem to suggest that the likely cause of premature failure of the beam plates

is ....

I believe this may be because ....

Additional information from the numerical data ....

....

Complete the list of key **health and safety**, **regulatory** and **sustainability** factors.

Health and safety

1 The chosen solution must pose no risk to the person fitting it to the joining beams. This applies to fitment when the beams are in position in the roof, or prior to fitment, on the ground.

....

....

Regulations

1 The chosen solution must be compatible with the relevant British standards relating to beam sizes

and load bearing. My research on materials and sizes of roof beams ....

....

....

Sustainability

1 The chosen solution must be manufactured from materials that can be recycled or reused at the

end of the product life cycle. Aluminium is suitable for recycling ....

....

....

Links To revise interpreting the brief, see pages 162–163 of the Revision Guide.

Revision activity 3: Produce a range of initial design ideas based on the client brief

Produce a range of (three or four) initial ideas based on the client brief, to include sketches and annotations.

Make sure that:
- your range of ideas are **appropriate** and **comprehensively** address the brief, including adaptations that are major improvements when compared to the existing product and the brief
- your ideas are communicated **clearly** and **concisely** with appropriate use of **annotation** and **technical terms**
- your ideas are **feasible**, **fit for purpose**, and reasonably different to the existing product shown and each other, when considering both form and approach.

⟩ **Guided** ⟩

 An extract from example idea 1 is sketched and annotated below. Use the following pages to sketch and annotate two of your own initial ideas. Use your research on **existing products** and **other aspects** of the design to **support**, **inform** and **inspire** your work.

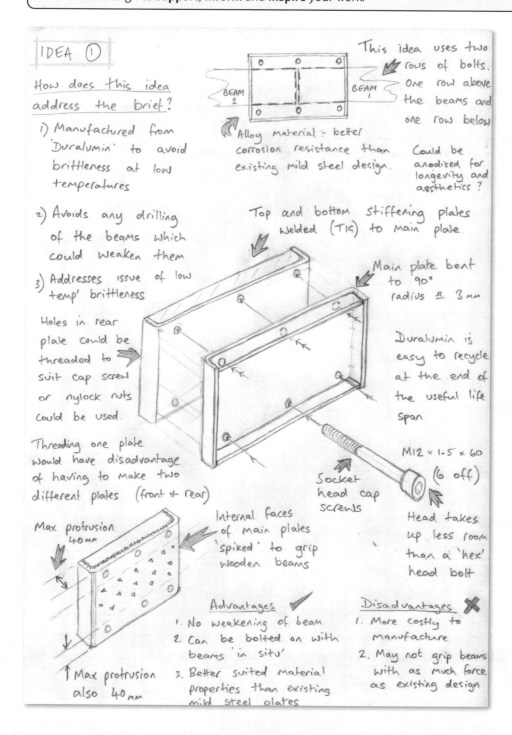

IDEA ①

How does this idea address the brief?

1) Manufactured from 'Duralumin' to avoid brittleness at low temperatures

2) Avoids any drilling of the beams which could weaken them

3) Addresses issue of low temp' brittleness

Holes in rear plate could be threaded to suit cap screw or nylock nuts could be used.

Threading one plate would have disadvantage of having to make two different plates (front + rear)

Max protrusion 40 mm

↑ Max protrusion also 40 mm

This idea uses two rows of bolts. One row above the beams and one row below

Alloy material = better corrosion resistance than existing mild steel design.

Could be anodised for longevity and aesthetics?

Top and bottom stiffening plates welded (TIG) to main plate.

Main plate bent to 90° radius ≅ 3 mm

Duralumin is easy to recycle at the end of the useful life span

M12 × 1.5 × 60 (6 off)

Socket head cap screws

Head takes up less room than a 'hex' head bolt

Internal faces of main plates 'spiked' to grip wooden beams

Advantages ✔
1. No weakening of beam
2. Can be bolted on with beams 'in situ'
3. Better suited material properties than existing mild steel plates

Disadvantages ✘
1. More costly to manufacture
2. May not grip beams with as much force as existing design

✏️ Sketch and annotate idea 2 below. Refer to the example on page 70, your research and the key headings on page 56.

Sketch and annotate idea 2 below. Refer to the example on page 70, your research and the key headings on page 56.

🔗 **Links** To revise initial ideas, see pages 164–165 of the Revision Guide.

Sketch and annotate **idea 3** below. Refer to the example on page 70, your research and the key headings on page 56.

Sketch and annotate idea 4 below. Refer to the example on page 70, your research and the key headings on page 56.

Revision activity 4: Develop a modified product proposal with relevant design documentation

Develop a modified product proposal with relevant design documentation. The proposal must include a solution, existing products, materials, manufacturing processes, sustainability, safety and other relevant factors.

Guided

> Your response to this activity will be significant in showing your skills and knowledge. Use the checklist below to make sure you **address the 10 important categories** as you **structure and document the development** of your final design. Use the following pages to **develop a modified product proposal with relevant design documentation for the brief** in this Workbook. Make **full use of the research** carried out on materials, processes and other aspects of the product and how it could be manufactured.

Optimisation

Prove that your solution is the best possible response to the original brief.

Justification

When compared with the existing product, prove that your solution is a significant improvement, with clear variation when considering form and approach.
Provide a balanced argument, giving reasons to support your view. Show how you arrived at your conclusions.

Informed decisions

Show you have a thorough understanding of existing alternative products and make it evident how the features of existing products relate to the chosen solution.

Materials

Prove you have investigated all possible options for materials and fully justified your choice.

Manufacturing

Prove you have investigated all possible options for methods of manufacture and fully justified your choice.

Sustainability

Show you have considered sustainability at all points of the product life cycle in relation to, for example, raw materials extraction, material production, production of parts, assembly, use and disposal/recycling in the context of the chosen solution.

Safety

Prove you have designed out all possible risk and that the proposal is safe to use and interact with, in relation to the existing product.

Documentation

Provide all the necessary information so that a third party can understand every aspect of your design, including, for example, a reasonably accurate orthographic projection.

Annotation

Provide detailed annotation so an engineer could manufacture the product.

Technical terminology

Use technical language correctly and accurately throughout, including annotation of your designs.

Guided

Use the sample **development response sheet 1 below** relating to manufacturing methods, to help you **develop** your own ideas into a **final solution**. Your final development section must cover: a developed solution, existing products, materials, manufacturing processes, sustainability, safety and any other relevant factors. This should be supported by any investigations you carried out into these areas in your research.

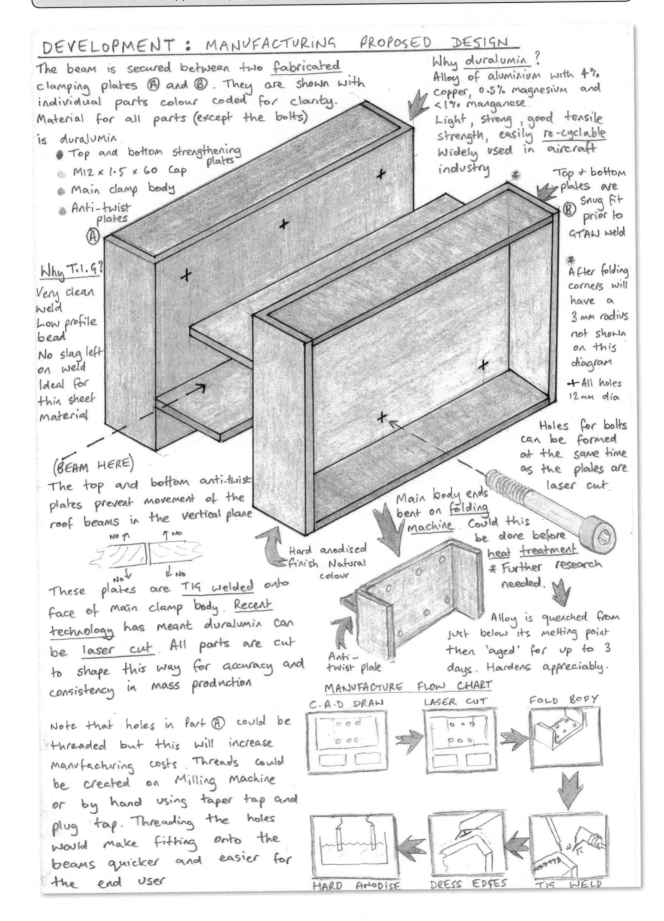

DEVELOPMENT : MANUFACTURING PROPOSED DESIGN

The beam is secured between two <u>fabricated</u> clamping plates Ⓐ and Ⓑ. They are shown with individual parts colour coded for clarity. Material for all parts (except the bolts) is duralumin.
- Top and bottom strengthening plates
- M12 × 1.5 × 60 Cap
- Main clamp body
- Anti-twist plates
Ⓐ

<u>Why T.I.G?</u>
Very clean weld
Low profile bead
No slag left on weld
Ideal for thin sheet material

(BEAM HERE)
The top and bottom anti-twist plates prevent movement of the roof beams in the vertical plane

These plates are TIG welded onto face of main clamp body. <u>Recent technology</u> has meant duralumin can be <u>laser cut</u>. All parts are cut to shape this way for accuracy and consistency in mass production

Note that holes in Part Ⓐ could be threaded but this will increase manufacturing costs. Threads could be created on Milling Machine or by hand using taper tap and plug tap. Threading the holes would make fitting onto the beams quicker and easier for the end user

Why <u>duralumin</u>?
Alloy of aluminium with 4% copper, 0.5% magnesium and <1% manganese.
Light, strong, good tensile strength, easily <u>re-cyclable</u> widely used in aircraft industry

Top & bottom plates are Ⓑ snug fit prior to GTAW weld

* After folding corners will have a 3 mm radius not shown on this diagram
+ All holes 12 mm dia

Holes for bolts can be formed at the same time as the plates are laser cut.

Main body ends bent on <u>folding machine</u>. Could this be done before heat <u>treatment</u>
* Further research needed.

Hard anodised finish. Natural colour

Anti-twist plate

Alloy is quenched from just below its melting point then 'aged' for up to 3 days. Hardens appreciably.

MANUFACTURE FLOW CHART
C.A.D DRAW → LASER CUT → FOLD BODY

HARD ANODISE ← DRESS EDGES ← TIG WELD

Develop your own ideas into a final solution using this page for **development response sheet 2**. Refer to the example on page 75, your research and the key headings on page 56.

To revise developing a modified product proposal into a final design, see pages 166–169 of the Revision Guide.

Links

Develop your own ideas into a final solution using this page for **development response sheet 3**. Refer to the example on page 75, your research and the key headings on page 56.

Develop your own ideas into a final solution using this page for **development response sheet 4**. Refer to the example on page 75, your research and the key headings on page 56.

Revision activity 5: Validate the design proposal

Your final Workbook entry must evaluate:

- success and limitations of the completed solutions
- indirect benefits and opportunities
- constraints
- opportunities for technology-led modifications.

> **To evaluate and validate your final design proposal you should:**
> - give a **balanced** and **thorough** appraisal of your design
> - provide a **sound rationale** for why the design solution is more **effective** in relation to the **brief**
> - **communicate**, with detailed **evidence**, how further **technology-led** modifications help optimise the solution.

  Read the **example validation** below, which relates to the sketched idea 1 on page 75, then write your **own validation** of your design proposal.

Validating the design proposal

The idea I have proposed is fabricated from Duralumin, which offers a number of advantages. Firstly, it performs well at low temperatures, which will help avoid the issues with the brittleness of the original plates. Secondly, Duralumin is very well suited to recycling at the end of the product life cycle.

One of the main features of this design is that there is no requirement to drill the beams, which can weaken them. The plates also feature a folded edge, which improves the stiffness of the joint. The success of this design is further enhanced through the use of threaded holes in one of the plates, which will reduce the time needed for assembly. This also cuts out the need for additional nuts and washers, which saves cost. Duralumin is a non-ferrous metal that does not require an applied finish, although anodising could be used to enhance the aesthetics and further extend the expected life cycle (cost would be increased).

The design does have some limitations. It is more difficult to manufacture than the original plates and may not grip the beams with as much force because the bolts are positioned above and below the faces of the beam. However, the inner faces do feature 'spikes' to grip the wooden beam, which will help eliminate any movement or 'creeping' of the beam over time.

This design directly addresses the following areas of the brief:

- The change in material has removed the chances of brittle failures in low temperature conditions.
- Using tapped holes in the plates instead of nuts and washers reduces the number of components in the kit.
- Not having to drill holes in the beam makes the joint easier to fit.
- By fitting the top bolts whilst on the floor the jointing system can be easily hooked over the beams, which will hold it in place during fixing. This makes it safer and easier to install than the original.

There is an indirect benefit from use of this design, too. These plates would be suitable to replace the existing plates without having to disturb the beams, or be used to repair beams that have developed weak spots.

The use of these beam plates may, however, be constrained in some places because they do protrude above and below the existing beam by 40 mm.

A future development of this design could be 'quick release' bolts that require only a half turn to lock into position. This would decrease fitment and removal time.

A technology-led adaptation for use in extreme environments would be to fit the jointing plates with low-cost sensors to monitor overloading or any movement in the beams. This would allow action to be taken before failure. This might be useful in extreme cold with high snow loading on the roof.

Validate your final design proposal. Use the headings to help you structure your validation.

Simple sketch of final proposal:

Brief explanation of main features:

....

....

....

....

....

....

....

....

Success of the completed solution:

....

....

....

....

....

....

....

....

Limitations of the completed solution:

....

....

....

....

....

....

....

....

Indirect benefits and opportunities of the completed solution:

....

....

....

....

....

....

....

....

Constraints of the completed solution:

....

....

....

....

....

....

....

....

Opportunities for technology-led modifications in relation to the completed solution:

....

....

....

....

....

....

....

....

Links To revise validation, see pages 150 and 170–171 of the Revision Guide.

Now **check** that your work covers all the **important aspects**. Below is a checklist for each activity and the **qualities** that you should show in your **solution**.

Tick off each one if you have addressed it during your completion of this Workbook revision task.

Activity criteria checklist

Activity	Breakdown of required tasks	Completed?
1	My Workbook activities are logical with an iterative approach throughout.	
	My design development links to research and requirements of the brief.	
	I have fully justified my design developments.	
	I have identified action points for the next session that are well-defined, logical and prioritised.	
2	I have interpreted the brief with a full list of product requirements.	
	I have accurately interpreted and calculated numerical data.	
	I have considered health and safety, sustainability and any relevant regulations with relevance to the context.	
3	I have given a range of ideas that fully address the requirements of the brief.	
	I have communicated each idea with clarity and technical annotation, linked to the brief.	
	Each idea is feasible, realistic, and fit for purpose.	
4	I have optimised my design solution and justified all alterations and developments.	
	My design proposal is informed and demonstrates a thorough understanding of existing products.	
	I have investigated material options and justified selection.	
	I have investigated manufacturing options and justified selection.	
	I have considered sustainability throughout life cycle of product.	
	I have clearly referenced the safety of the design and designing out risks.	
	My formal documents allow a third party to make the product correctly, with concise annotation of the solution and use of accurate technical terminology.	
5	I have shown a balanced and thorough appraisal of success and limitations of completed solutions, indirect benefits and opportunities, and constraints.	
	I have provided a sound rationale that states why the design is more effective in relation to the brief than the original solution.	
	I have communicated with detailed evidence how technology-led modifications could further optimise my chosen solution.	

END OF TASK

Unit 6: Microcontroller Systems for Engineers

Your set task

Unit 6 will be assessed through a task, which will be set by Pearson. You will need to complete a practical task where you develop a prototype microcontroller system to solve a problem.

Your Revision Workbook

This Workbook is designed to **revise the skills** that will be needed in your assessed task. The selected content, outcomes, questions and answers are provided to help you revise content and ways of applying your skills. Ask your tutor or check the Pearson website for the most up-to-date Sample Assessment Material and Mark Scheme to get an indication of the structure of your actual assessed task and what this requires of you. The detail of the actual assessed task may change so always make sure you are up to date. Make sure you check the instructions in relation to completing the assessment on computer using the appropriate hardware and software as listed in the unit specification, the use of a calculator, and the requirements and format when submitting evidence. The table below indicates the only acceptable microcontroller hardware and programming languages for Unit 6.

Hardware device family	Software Integrated Development Environment (IDE)	Programming language
Arduino™/ Genuino™	Arduino™ IDE or Flowcode	Arduino™ C or Flowchart
PIC®	MPLAB® IDE (MPLAB® C) or Flowcode	C or Flowchart
PICAXE®	PICAXE® Editor	BASIC or Flowchart
GENIE®	GENIE® Studio or Circuit Wizard	BASIC or Flowchart

To support your revision, this Workbook contains a revision task to help you revise the skills that might be needed in your assessed task.

Responding to activities

Your response to the activities will involve you in reading and understanding a scenario and client brief, followed by:

- Task planning and system design changes
- Analysis of the brief
- System design
- System assembly and programming
- System testing and result analysis
- System shown in operation.

The development of the system is an **iterative process**, with activities in each stage combining with each other, resulting in the system in operation.

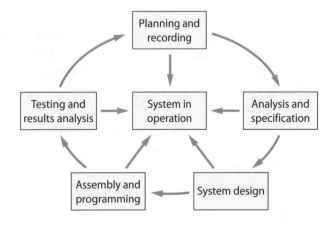

🔗 **Links** To help you revise the skills for your Unit 6 set task, this Workbook contains a full revision task starting on page 84. See the introduction on page iii for more information on features included to help you revise.

Revision task brief

To support your revision, this Workbook contains a full revision task to help you revise the skills that might be needed in your assessed task. The details of the actual assessed task may change so always make sure you are up to date. Ask your tutor or check the Pearson website for the most up-to-date Sample Assessment Material to get an idea of the structure of your assessed task and what this requires of you.

In this Workbook, examples of developing a solution in response to the brief are given using selected hardware and software. If you are developing your solution using a different microcontroller platform, you should follow the same approach. Take the software instructions and apply them to those that perform the same functions in the microcontroller platform that you are using.

Where tasks require completion on computer, you can print out your solutions and keep them as a record in this Revision Workbook.

 Start by reading through the **task briefs** that follow, underlining key information. Then read through the Revision task on page 85.

Scenario

You are employed as a technician in the biological science department of a university. A series of experiments are to be conducted that investigate the growth response of seedlings to light. Fast-growing seeds, such as *Brassica rapa* (fast plants), *Sinapis alba* (white mustard), *Raphanus sativus* (radish), will be used for the experiments.

You have been asked to produce some of the equipment that will be used in the experiments. You have previously used microcontroller-based systems for other projects within the department successfully, which is why you have been selected to design and implement equipment for this experiment.

Client brief

The scientists want to determine the effects of several factors. They want to observe if the colour of the light is an important factor in the growth of plants. They have decided that they want to compare how seedlings react to red, green and white light. Another factor they believe may influence the seedlings' growth is the intensity of the light.

In order to test the response of the seeds to the colour of light the experiment requires that a seed will be exposed to a repeating on/off sequence of the different coloured lights. Each light should be on for a user-selectable duration of between 1 and 60 seconds. It is expected that the experiment may take up to 10 days to show measurable results. During this period the seeds will be excluded from ambient light sources.

In order to test the response of the seeds to the intensity of light each light should be able to have its output varied in increments of 25%. Each light should emit the same intensity of light during a single experiment.

Each of the light sources should be 50 mm above the seed, 15 mm from the centre of the seed. Each light source should be separated by 120°. The light sources should not affect the environmental temperature of the experiments.

You have been allocated a budget of £10.00, excluding the microcontroller hardware itself.

Client name: Kennett University: Science Department
- Task planning and system design changes made during the development process
- Interpret a brief into a technical specification with operational requirements
- Design a test plan based on operational requirements
- Select and describe appropriate input/output components and how they will work together, giving details and justification with system connection diagrams/schematics
- Design the program structure
- Produce a functional system, with annotated copy of all the code to demonstrate understanding
- Test the system and analyse the outcomes from testing
- Produce an audiovisual recording of the system in operation, to the specified length.

Revision task

 Read through the revision task and activities, underlining key information. Then complete the response on page 86.

Task

Design, assemble, program and test equipment to be used in experiments that investigate the growth response of fast-growing seedlings, following appropriate development processes and using a microcontroller. You will need to complete your prototype test equipment (including testing and documentation) in stages, and the activities below will help you to structure your development work.

Revision activity 1: Task planning and system design changes (see pages 87–100)

At the start of your task create a short project time plan in your Revision Workbook and use it to monitor your progress. Also, during the development process use the Workbook to record changes to your original designs, providing details of any issues encountered and solutions discovered, with a justification. You will also need to log actions planned for each next session, considering what needs to be done, what the priorities are, what resources are needed, and how you will determine if you have successfully achieved what you set out to do.

Revision activity 2: Analysis of the brief (see pages 101–106)

Analyse the scenario, background information and product requirements from the client brief in your Workbook and use this to generate a technical specification for the system. Your technical specification should then be used to help complete a test plan to test the suitability of the final solution.

Revision activity 3: System design (see pages 107–112)

Select and justify input and output devices and formulate an initial design for the system in your Workbook. You should detail any interfacing/design requirements that you require. Create an outline plan for the program structure based on your hardware selection and system design.

Revision activity 4: System assembly and programming (see pages 113–121)

Assemble your hardware, author your program and annotate your code. Insert the annotated code into your Workbook.

Revision activity 5: System testing and results analysis (see pages 122–131)

Test your system against your test plan (from Revision activity 2) and record the outcome of each test, using the template in your Workbook. Analyse your test results and evaluate your system for conformance against technical specification (and, hence, the client's brief).

Revision activity 6: System in operation (see pages 132–141)

Make an evidential audiovisual recording that demonstrates your solution in operation, with commentary that explains the operation of the system, and how its behaviour is linked with your chosen hardware and program. The audiovisual recording of the system in operation should also provide visual evidence of the outcome from most of the tests (see Revision activity 5).

Understand the scenario, client brief and task

Read the **revision task brief** and **activities** thoroughly to get a picture of what you are being asked to do. You need to ensure you provide evidence of meeting all the requirements included in the brief. If you miss relevant information at this stage, it will reduce your ability to provide a complete solution.

As you read the brief, you will **develop an initial concept** of what the task will involve and this concept will be refined as you progress. At this stage you also need to start formulating questions. The **answers** to these **questions** will allow you to develop the **technical specification**.

Guided Use the key pieces of information from the **revision scenario**, **client brief** and **task** that you need to consider in your solution, to answer the questions that follow.

Key questions on the client brief

1 What inputs are required?

    ....

    ....

2 What function are the inputs required to perform?

    ....

    ....

3 What outputs are required?

    ....

    ....

4 What functions are the outputs required to perform?

    ....

    ....

5 What aspects of the client brief relate to user experience?

    ....

    ....

6 What constraints are imposed?

    ....

    ....

Revision activity 1: Task planning and system design changes

To start the task create a short project time plan and use it to monitor your progress. Also, during the iterative development process use the logs on pages 89–100 to record changes to your original designs, providing details of any issues encountered and solutions discovered, with a justification. You will also need to log actions planned for each next session, considering what needs to be done, what the priorities are, what resources are needed, and how you will determine if you have successfully achieved what you set out to do.

You should:
- Demonstrate a **structured** approach to the **iterative development process**.
- Carry out the **development activities** in an appropriate **order**.
- Use **accurate** technical terminology.
- **Justify** any **changes** made to your original designs, demonstrating **logical** chains of reasoning between testing and the changes made.
- Plan well-defined and logical **action points** for each session.
- Complete a **log book record** for each **session**.

Project time plan

To complete your initial task plan you should:
- Produce a personal timetable that sets critical dates and times by which you need to have completed each task. If an individual task is not complete you may still need to move on so that you meet the overall deadline.
- Consider what the task requires and break down your plan further in a structured way.

> **Guided** > Use the template below to produce a **timetable** to complete the revision task on pages 84–85. For your actual assessment, check with your tutor or the Sample Assessment Material on the Pearson website so you know how much time you will have to complete a task.

Start date of task	Date.......... Month.......... Year..........		
Activity	Allocation of time	Date to be completed	Time to be completed
(1) Task planning and system design changes			
(2) Analysis of the brief			
(3) System design			
(4) System assembly and programming			
(5) System testing and result analysis			
(6) System in operation			
Total			

To break down your initial task plan further, you do not have to complete it in chronological order. You could analyse the client brief, part of Activity 2, so you have a better understanding of what is required. You will then be able to plan and monitor your time in a structured and ordered way.

Complete the plan below with the **key tasks** required of you, overall. Add as many entries as you think are required. Remember that this is an **initial** task plan and can be modified throughout the task if and when required.

Initial task plan

1. Read through the task to understand what is required.
2. Note down the sessions needed to complete the work.
3. Analyse the scenario and client brief.
4. Produce the initial project time plan.
5. At the start of each session review plan from previous one.

....

....

....

....

....

....

....

....

....

....

....

....

....

....

....

....

....

....

....

....

....

....

....

....

Task session log book tables

 During each **session** in your task you need to **complete a log book table** using the entry format below to:

- **record your progress** and **solutions** to issues encountered, with **justification**
- **plan** what you want to achieve next time.

Twelve log book tables are provided on the following pages. Use as many as you need for your sessions. You can create more if needed. Start by completing a log of your first session, then go to page 102 to analyse and interpret the brief.

Log book record: Session 1

Date:	
You may also find it useful to include times.	

General comments: ....

....

....

....

....

This should be an overview of what you planned to complete and what was actually achieved.

Issues encountered and solutions, with justification: ....

....

....

....

....

....

....

This should include: what the problem was, the cause of the problem, how you resolved the problem, why you chose your solution rather than an alternative.

Action list for the next session: ....

....

....

....

....

....

You should consider what needs to be done, what the priorities are, what resources are needed, how you will determine if you have successfully achieved what you set out to do.

Log book record: Session 2

Guided

Date:

> You may also find it useful to include times.

General comments: ....

....

....

....

....

> This should be an overview of what you planned to complete and what was actually achieved.

Issues encountered and solutions, with justification: ....

....

....

....

....

....

....

> This should include: what the problem was, the cause of the problem, how you resolved the problem, why you chose your solution rather than an alternative.

Action list for the next session: ....

....

....

....

....

....

> You should consider what needs to be done, what the priorities are, what resources are needed, how you will determine if you have successfully achieved what you set out to do.

 See pages 229–231 of the Revision Guide for guidance on monitoring your progress, which would help inform a log book entry, and pages 113–118 of the Revision Workbook for examples.

Log book record: Session 3

Date:

You may also find it useful to include times.

General comments: ....

....

....

....

....

This should be an overview of what you planned to complete and what was actually achieved.

Issues encountered and solutions, with justification: ....

....

....

....

....

....

....

This should include: what the problem was, the cause of the problem, how you resolved the problem, why you chose your solution rather than an alternative.

Action list for the next session: ....

....

....

....

....

....

You should consider what needs to be done, what the priorities are, what resources are needed, how you will determine if you have successfully achieved what you set out to do.

Log book record: Session 4

Date:

You may also find it useful to include times.

General comments: ....

....

....

....

....

This should be an overview of what you planned to complete and what was actually achieved.

Issues encountered and solutions, with justification: ....

....

....

....

....

....

....

This should include: what the problem was, the cause of the problem, how you resolved the problem, why you chose your solution rather than an alternative.

Action list for the next session: ....

....

....

....

....

....

You should consider what needs to be done, what the priorities are, what resources are needed, how you will determine if you have successfully achieved what you set out to do.

Log book record: Session 5

Date:

> You may also find it useful to include times.

General comments: ....
....
....
....
....

> This should be an overview of what you planned to complete and what was actually achieved.

Issues encountered and solutions, with justification: ....
....
....
....
....
....
....

> This should include: what the problem was, the cause of the problem, how you resolved the problem, why you chose your solution rather than an alternative.

Action list for the next session: ....
....
....
....
....
....

> You should consider what needs to be done, what the priorities are, what resources are needed, how you will determine if you have successfully achieved what you set out to do.

Log book record: Session 6

> **Guided**

Date:

> You may also find it useful to include times.

General comments: ....

....

....

....

....

> This should be an overview of what you planned to complete and what was actually achieved.

Issues encountered and solutions, with justification: ....

....

....

....

....

....

....

> This should include: what the problem was, the cause of the problem, how you resolved the problem, why you chose your solution rather than an alternative.

Action list for the next session: ....

....

....

....

....

....

> You should consider what needs to be done, what the priorities are, what resources are needed, how you will determine if you have successfully achieved what you set out to do.

Log book record: Session 7

Date:

You may also find it useful to include times.

General comments: ....
....
....
....
....

This should be an overview of what you planned to complete and what was actually achieved.

Issues encountered and solutions, with justification: ....
....
....
....
....
....
....

This should include: what the problem was, the cause of the problem, how you resolved the problem, why you chose your solution rather than an alternative.

Action list for the next session: ....
....
....
....
....
....

You should consider what needs to be done, what the priorities are, what resources are needed, how you will determine if you have successfully achieved what you set out to do.

Log book record: Session 8

> Guided

Date:

> You may also find it useful to include times.

General comments: ....
....
....
....
....

> This should be an overview of what you planned to complete and what was actually achieved.

Issues encountered and solutions, with justification: ....
....
....
....
....
....
....

> This should include: what the problem was, the cause of the problem, how you resolved the problem, why you chose your solution rather than an alternative.

Action list for the next session: ....
....
....
....
....
....

> You should consider what needs to be done, what the priorities are, what resources are needed, how you will determine if you have successfully achieved what you set out to do.

Log book record: Session 9

Guided

Date:

You may also find it useful to include times.

General comments: ....
....
....
....
....

This should be an overview of what you planned to complete and what was actually achieved.

Issues encountered and solutions, with justification: ....
....
....
....
....
....
....

This should include: what the problem was, the cause of the problem, how you resolved the problem, why you chose your solution rather than an alternative.

Action list for the next session: ....
....
....
....
....
....

You should consider what needs to be done, what the priorities are, what resources are needed, how you will determine if you have successfully achieved what you set out to do.

Log book record: Session 10

> Guided

Date:

> You may also find it useful to include times.

General comments: ....

....

....

....

....

> This should be an overview of what you planned to complete and what was actually achieved.

Issues encountered and solutions, with justification: ....

....

....

....

....

....

....

> This should include: what the problem was, the cause of the problem, how you resolved the problem, why you chose your solution rather than an alternative.

Action list for the next session: ....

....

....

....

....

....

> You should consider what needs to be done, what the priorities are, what resources are needed, how you will determine if you have successfully achieved what you set out to do.

Log book record: Session 11

Date:

> You may also find it useful to include times.

General comments: ....
....
....
....
....

> This should be an overview of what you planned to complete and what was actually achieved.

Issues encountered and solutions, with justification: ....
....
....
....
....
....
....

> This should include: what the problem was, the cause of the problem, how you resolved the problem, why you chose your solution rather than an alternative.

Action list for the next session: ....
....
....
....
....
....

> You should consider what needs to be done, what the priorities are, what resources are needed, how you will determine if you have successfully achieved what you set out to do.

Log book record: Session 12

> Guided

Date:

> You may also find it useful to include times.

General comments: ....

....

....

....

....

> This should be an overview of what you planned to complete and what was actually achieved.

Issues encountered and solutions, with justification: ....

....

....

....

....

....

....

> This should include: what the problem was, the cause of the problem, how you resolved the problem, why you chose your solution rather than an alternative.

Action list for the next session: ....

....

....

....

....

....

> You should consider what needs to be done, what the priorities are, what resources are needed, how you will determine if you have successfully achieved what you set out to do.

Revision activity 2: Analysis of the brief

Analyse the scenario, background information and product requirements from the client brief and task (pages 84–85) and use this to generate a technical specification. Your technical specification should then be used to help complete a test plan to test the suitability of the final solution.

> **You should:**
> • Interpret the brief into a comprehensive set of **operational requirements** that fully meets the brief and considers **enhanced user experience**.
> • Produce a comprehensive **test plan** with parameters that are designed to confirm a fully **functioning system** including **unexpected events**.
> • Complete a **log book record** for each **session** with **actions** for the next session (see pages 89–100).

Identify the key requirements your solution must meet

> For Revision activity 2, **analysis of the client brief** leads to the **production of a technical specification**. A useful approach is to **extract** the **key requirements** from the client brief to make what is required clearer. Some of the information provided in the scenario / client brief will be background information and will not have a direct influence on what you are required to produce.

> Use the space below to identify **9 key requirements** your solution must meet. Three have been identified for you.

Key requirements are:

1. Red, green and white lights

2. User-selectable duration of between 1 and 60 seconds

3. £10.00 budget for components

4. ....

5. ....

6. ....

7. ....

8. ....

9. ....

Interpret the brief into operational requirements

When you have identified the key criteria your solution must meet, you can interpret the brief into a set of operational requirements. **Operational requirements** means things that the system must be able to perform while in use. When you specify your operational requirements make sure that you also comment on why you have included them. This helps determine if they are cohesive. Remember to consider enhanced user experience.

> **Guided**

The table below presents several operational requirements identified for the given client brief.
- The **first entry** should be completed to suit the **development platform** you intend to use e.g. Arduino™ Uno and Flowcode).
- You should also identify the **types of component** you will use in your system.

Extend the table to cover all operational requirements you think will be important.

Operational requirement	Why this is important
The system will be programmed onto .... using	This is the system I am familiar with and have the skills to use.
The system will be powered by a power supply not a battery pack.	The system may have to run for 10 days. A battery pack could run out before that time is up.
The system will use LEDs as the light source.	These are available in the colours required and do not generate heat. This means they won't make the environment hotter.
The user must be able to select times between 1 and 60 whole seconds.	The client brief is not specific about the requirement for fractions of a second. To make every possible value available would probably be too complex to achieve.

Operational requirement	Why this is important
....	....
....	....
....	....
....	....
....	....
....	....
....	....
....	....
....	....

Produce a test plan

Having determined the **operational requirements**, the next stage is to determine if these have **been met**.

The test plan needs to include **details of the parameters** that will be used to judge whether the system conforms with the needs of the client brief. There should be a link between **operational requirements** and the **test plans**. Remember to consider **unexpected events**.

When writing the test plan, consider how you will **undertake** the **test**.

 Guided

The table below describes several tests that could be completed to **test the performance** of the system. Extend the table to cover all those you think would be important.

Description of test	How this could be achieved
The method used for setting the time the LEDs are on should be operated to ensure that all numbers between 1 and 60 are possible.	The system could be adjusted one step at a time from 1 to 60 and the results observed and recorded.
The period of time the lights stay on for should match that set by the user.	Time using a stopwatch or oscilloscope. As it would take too long to test all 60 values, performance at 1, 30 and 60 seconds could be measured. If these are accurate, the others should be.

Record the outcome of your results

Guided

 As part of your test plan you need to **record** the outcomes of your tests using the **test plan template** below. Having entered the results, you will need to **analyse** them.

Test	Purpose of test	Test data	Expected result	Actual result	Comments and justification
1	To establish a method to illuminate the LEDs for between 1 and 60 seconds	1 s = 30 s = 60 s =	Length of time LEDs are lit when system operates from 1 to 60 seconds 1 s = 1 30 s = 30 60 s = 60		

Analysis of results

(How far the results show that the system meets the client brief)

....

....

....

....

....

....

....

....

....

....

....

....

....

....

....

....

....

....

....

....

....

....

....

....

....

....

....

....

 Links To revise technical specifications and test plans, see page 233 of the Revision Guide.

Revision activity 3: System design

Select and justify input and output devices and formulate an initial design for the system. You should detail any interfacing/design requirements that you require. Create an outline plan for the program structure based on your hardware selection and system design.

> **You should:**
> - Select **input and output devices** appropriate for the **operational requirements**.
> - Produce **diagrams** and/or **written commentary** that describe accurately and in detail how the input and output devices interface with the microcontroller.
> - Use **technical terminology** and **industry standard conventions** appropriately throughout.
> - Design a **program structure** that breaks down key operations into relevant constructs that **link** logically and coherently, including the handling of **unexpected events**.
> - Complete a **log book record** for each **session** with **actions** for the next session (see pages 89–100).

Select appropriate hardware and determine interface with the microcontroller

> Having analysed and interpreted the brief into operational requirements and developed a test plan, the next stage is to select appropriate hardware and determine how it will interface with the microcontroller. As the microcontroller itself is the centre of the system, it is logical to consider what capability is provided by the microcontrollers you have access to and determine if they will meet the demands of the client brief.

> **Guided**

> Obtain **a pin out diagram** for the microcontroller you plan to use. Below is an example for the ATMEGA328P that is used in an Arduino™ controller. Use the examples that follow for Revision activity 3 to apply to your own project.

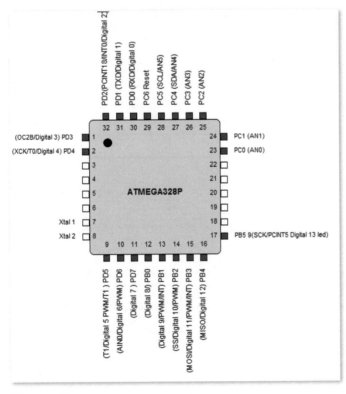

Microcontroller connection table

Pin	Part	Function 1	Function 2	Connected to?
1	D3	Digital 3	Interrupt 1	
2	D4	Digital 5		
9	D5	Digital 5	PWM	

Pin out diagram

Develop your solution and connection table

When you have a pin out diagram for the microcontroller you will use, you need to **develop your solution** for the task and produce an appropriate **connection table** showing **pins**, **ports** and **functions**. Some pins will have more than two functions, but adding more than two to your table may cause confusion. In the table below, the 'connected to' column will be used when you plan how to interface the input and output devices.

> **Guided** Complete the **connection table** below with the **solution** for your task.

Microcontroller connection table

Pin	Port	Function 1	Function 2	Connected to?
1				
2				
9				
10				
11				
12				
13				
14				
15				
16				
17				
23				
24				
25				
26				
27				
28				
29				
30				
31				
32				

Select specific components

When you have identified the generic types of components that will provide the input and output hardware for your system, in the 'connected to' column on the previous page, you need to **refine** this to indicate **specific components**. Ensure you know how to find the **type of components** you will need efficiently and how to specify the ones you want to use; for example, using a component catalogue. You are expected to use **industry standard conventions**. Using a standard template for ordering components could be one way to evidence this.

Manufacture Brighto
Standard 5 mm LED
A range of low cost general purpose LEDs suitable for a wide range of applications

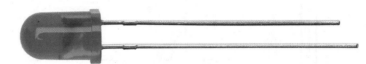

Tech Spec

Colour	Wavelength	Brightness	Max current	Max voltage	Order code
Red	700 mm	10 mcd	30 mA	2.8 V	LEDR-048
Yellow	590 mm	15 mcd	30 mA	2.8 V	LEDY-049
Green	570 mm	20 mcd	30 mA	2.8 V	LEDG-050

Price per pack of 100
Order code

LEDR-048	£5.40
LEDY-049	£6.70
LEDG-050	£6.70

The image shows a typical section from a supplier's printed component catalogue.
Frequently there will be many variations of the same product.

Use the sources of information available to you to **identify** and **select** a suitable **supplier**, or suppliers, for the components listed below. Extend the table to add the other components you have identified as needed to complete your system.

Component required	Number required	Supplier	Part number	Cost
16 * 2 LCD screen	1			
Rotary switch 2 pole 6 way	1			

Create a schematic diagram

You will need to describe how the **hardware** components of your system will **interface** with the **microcontroller** you are using, as shown in the example below.

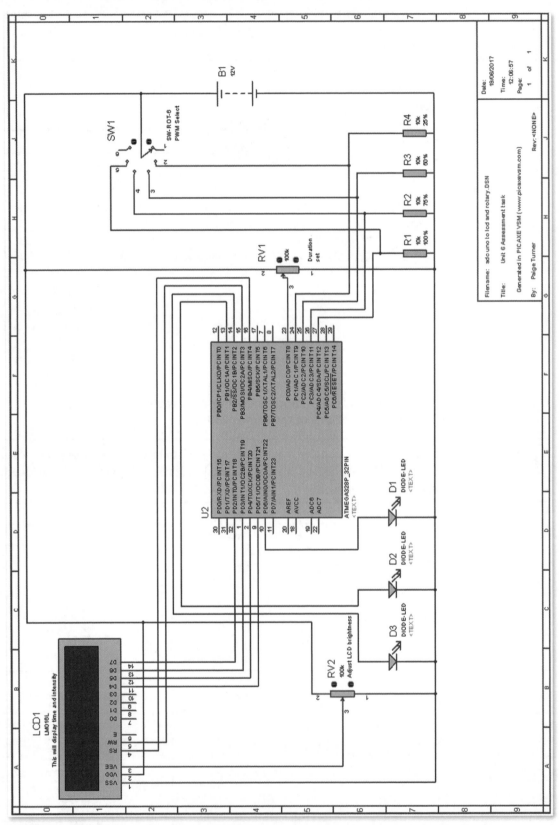

Schematic diagram of a typical industry standard generated for an Arduino™ Uno chip completed using PICAXE® VSM software.

Using a method of your choice, reproduce a version of a **schematic diagram** below that uses the **microcontroller** you would use to develop the system and **input/output devices** of your choice. The process of developing your system will be **iterative**. You might need to amend your original circuit diagram several times when completing the task.

My schematic diagram

Create an outline plan for the program structure

Having identified the components that will make up the hardware elements of the system, you need to plan the **software system** that will control them. The plan will need to take into account the types of hardware you want to use for your system. This is an **iterative** process – you may make decisions about software that will require you to select different hardware. The table below shows a possible **method** of **planning** the system.

Guided Complete the table to **plan** how you would **implement a system** to meet the **client brief**. Example entries are below. You need to complete the table with entries for the hardware you will use.

Action	Hardware	Coding construct
User sets time for LEDs to be on	Potentiometer	Analogue to digital conversion
User sets intensity of LEDs	Rotary switch	Test if the input pin the switch is connected to is high or low
The system provides the user with feedback on the values they have selected	LCD	Print to LCD

Action	Hardware	Coding construct

Links To revise input/output devices and program structure, see pages 178–190, 206–209 and 235–236 the Revision Guide.

Revision activity 4: System assembly and programming

Assemble your hardware, author your program and annotate your code. Insert the annotated code into your Workbook.

> **You should:**
> - Select and use a **range of appropriate constructs** correctly. The program should be **concise, efficient** and be able to handle some **unexpected events**.
> - Annotate the code to demonstrate thorough understanding of the **key areas of the program** and **underpinning constructs**.
> - **Organise, structure** and **format** the program so that a **competent third party** could interpret and update it.
> - Complete a **log book record** for each **session**, with actions for the next session (see pages 89–100).

Assemble, program and make records of your system

As you **assemble** and **program** your system, you need to **record** how you have developed your solution.

Guided example

 Read through some **example log book entries** on pages 113–118. Then use the guidance on pages 119–121 to apply them to your **own project** for **system assembly** and **records** for Revision activity 4.

Date: 22 June 2017 14:00	
General comments:	
Today I started to write the software and develop the hardware to use the ADC and display the value on the LCD. I simulated them to check they worked.	
Issues encountered and solutions, with justification:	
I had some problems with getting the settings right to display the characters where I wanted them to be. To get them in the correct place I just experimented by changing their values. This approach was the quickest way.	
The reading from the ADC gave a value of 255 max. I wanted it to be 60. I changed the value by dividing the reading by 4.2, using a maths function. 255/4.2 = 60.7, which is displayed as 60.	
Action list for the next session:	
I will produce a system diagram to show the connections and assemble the hardware to test if it works.	

Guided example

This is the first flowchart produced:

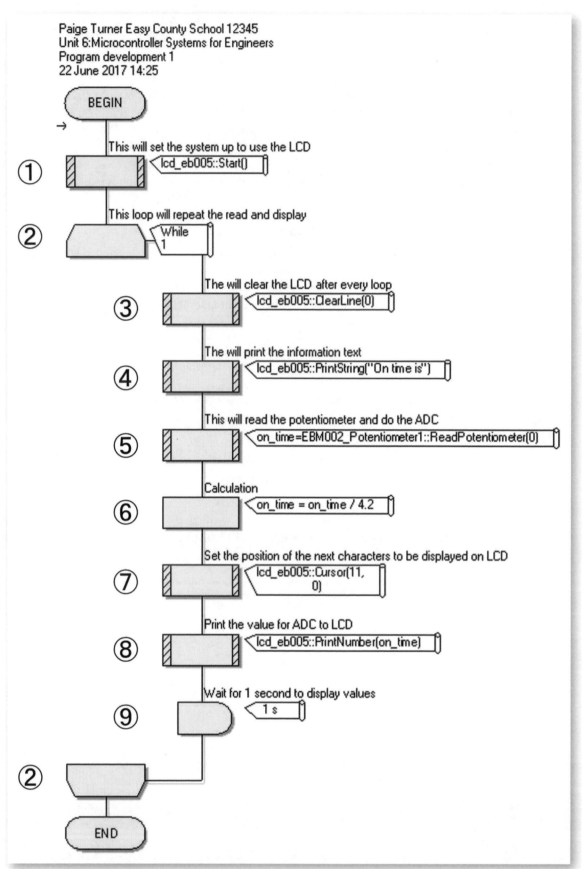

Paige Turner Easy County School 12345
Unit 6:Microcontroller Systems for Engineers
Program development 1
22 June 2017 14:25

BEGIN

① This will set the system up to use the LCD
lcd_eb005::Start()

② This loop will repeat the read and display
While 1

③ The will clear the LCD after every loop
lcd_eb005::ClearLine(0)

④ The will print the information text
lcd_eb005::PrintString("On time is")

⑤ This will read the potentiometer and do the ADC
on_time=EBM002_Potentiometer1::ReadPotentiometer(0)

⑥ Calculation
on_time = on_time / 4.2

⑦ Set the position of the next characters to be displayed on LCD
lcd_eb005::Cursor(11, 0)

⑧ Print the value for ADC to LCD
lcd_eb005::PrintNumber(on_time)

⑨ Wait for 1 second to display values
1 s

②

END

This method of recording the development of the program shows a **structured** approach being carried out in an **appropriate** order.

The flowchart has been **annotated** such that a third party could **interpret** it.

This is the result of the simulation:

I added the calculation to change the maximum value to 60.

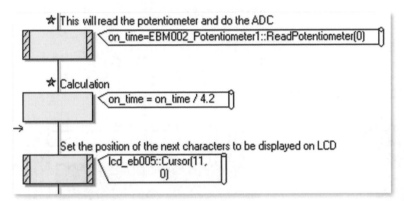

This is the result of the simulation:

This evidence **supports** and **justifies** the changes made.

Date: 23 June 2017 15:30	
General comments:	
I built and tested the hardware to read the ADC and print the reading to the LCD.	
Issues encountered and solutions with justification:	
When I first downloaded the program to the microcontroller the message did not display as expected. I carefully checked the connections and realised that I had swapped over connections 11 and 12 on the board. Swapping them to the correct positions solved the problem. This was the only solution that would work.	
Action list for the next session:	
I will produce a system diagram to show the connections and assemble the hardware to test if it works.	

Guided example

This is the circuit diagram I produced using Fritzing. I have also written the connections.

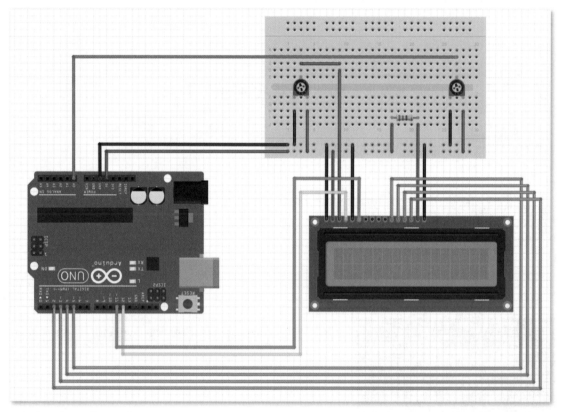

Fritzing breadboard

Text version of connections:

LCD	VSS	VDD	VO	RS	RW	E	D4	D5	D6	D7	A	K
Uno label	GND	5V	5V	12	GND	11	5	4	3	2	5V	GND
Uno port				84		83	D5	D4	D3	D2		

From this I obtained this schematic diagram:

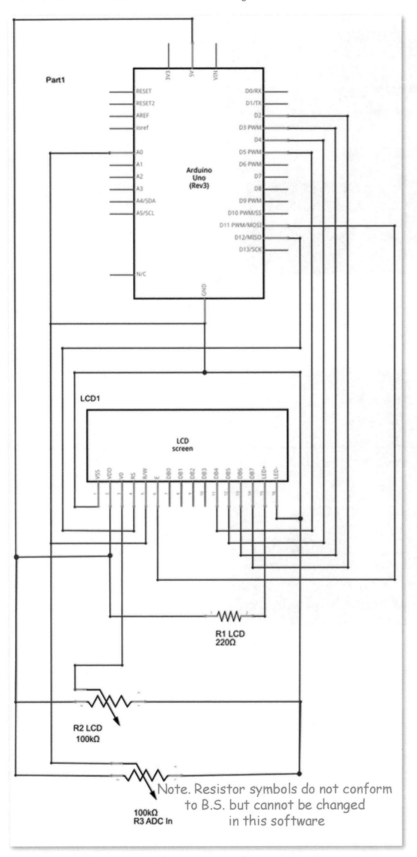

Note. Resistor symbols do not conform to B.S. but cannot be changed in this software

In this diagram, **industry standard conventions** have not been used for the resistor symbols. If **limitations of software** being used do not allow you to comply with expected standards, you should **annotate** your **evidence** to indicate you are aware of this.

Guided
example

This shows a problem with the display on the LCD. I had connected the wires incorrectly:

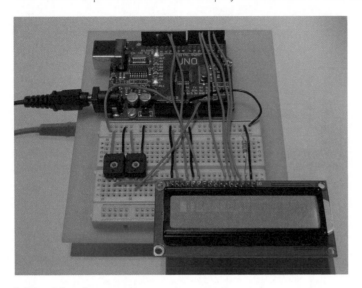

LCD wiring incorrect

This shows the problem corrected:

LCD wiring correct

Produce your own version of the system

Pages 113–118 show how the **hardware** and **software** converts the reading from the **potentiometer** to a value displayed on the **LCD**. You will now use these pages as a guide to producing your own version of the system.

 Using a **development platform** of your choice (i.e. **microcontroller** and associated **software**), repeat the process shown in the example to develop the **same working system**.

In order to complete this you should:

1 Write a functioning **program** to use an analogue to digital convertor (ADC).

2 Scale the **output** of the ADC such that the range is 0–60.

3 Extend this program to **display** the scaled ADC value on a LCD. Remember to **annotate** your **code** so that a third party can interpret it – a guide is provided below.

4 Simulate the **operation** of the **system**.

5 Produce a **circuit diagram** for the **hardware microcontroller interface**.

6 Construct the **system**.

7 Download the program to **microcontroller**.

8 Check the **system operates** as expected.

In order to complete the tasks above:

- You will need to access information to help make the appropriate connections between your selected hardware and the microcontroller. If you do not have access to some of the hardware, such as the LCD, you should substitute an equivalent method of displaying the information.

- It will probably be useful to use help files in the software you are using to look up the equivalent instructions to those used in the flowchart.

- If you do not have access to software to produce the connections diagrams, they can be produced manually.

- See page 120 for an example LCD flowchart to guide your own version of the program in a programming language of your choice.

- Use page 121 to place a printout of your program and response to the tasks above, as a record.

Guided example This is an **example of a program** to read an ADC and display the values on an LCD. It can be used as a guide to help you produce your **own version of the program** in a programming language of your choice.

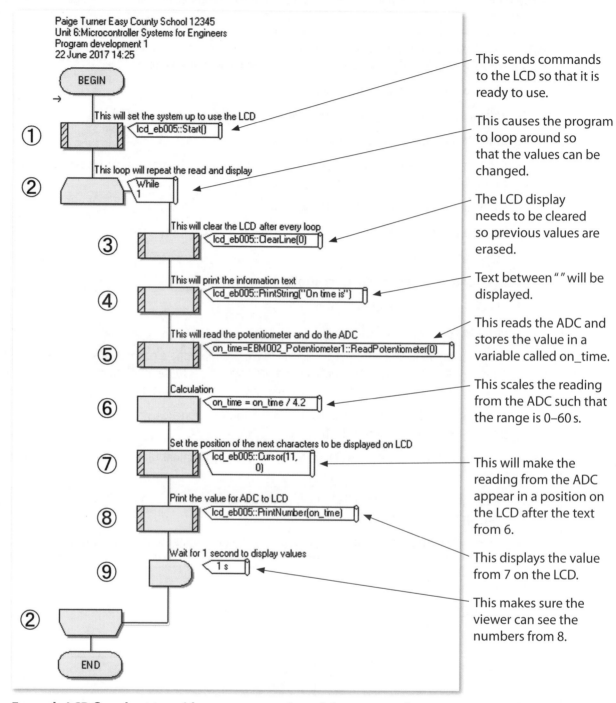

Paige Turner Easy County School 12345
Unit 6:Microcontroller Systems for Engineers
Program development 1
22 June 2017 14:25

BEGIN

① This will set the system up to use the LCD
lcd_eb005::Start()

② This loop will repeat the read and display
While 1

③ This will clear the LCD after every loop
lcd_eb005::ClearLine(0)

④ This will print the information text
lcd_eb005::PrintString("On time is")

⑤ This will read the potentiometer and do the ADC
on_time=EBM002_Potentiometer1::ReadPotentiometer(0)

⑥ Calculation
on_time = on_time / 4.2

⑦ Set the position of the next characters to be displayed on LCD
lcd_eb005::Cursor(11, 0)

⑧ Print the value for ADC to LCD
lcd_eb005::PrintNumber(on_time)

⑨ Wait for 1 second to display values
1 s

②

END

This sends commands to the LCD so that it is ready to use.

This causes the program to loop around so that the values can be changed.

The LCD display needs to be cleared so previous values are erased.

Text between " " will be displayed.

This reads the ADC and stores the value in a variable called on_time.

This scales the reading from the ADC such that the range is 0–60 s.

This will make the reading from the ADC appear in a position on the LCD after the text from 6.

This displays the value from 7 on the LCD.

This makes sure the viewer can see the numbers from 8.

Example LCD flowchart to guide your own version of the program in a programming language of your choice.

A solution in your chosen programming language

Produce your **own solution** in a **programming language of your choice**, using the program on page 120 to guide you. Print it out and paste it below as a record, along with your responses to the task on page 119.

 Links To revise flowcharts, see page 237 of the Revision Guide.

Revision activity 5: System testing and result analysis

Test your system against your test plan (from Revision activity 2) and record the outcome of each test, using the template in your Workbook (page 105).

Analyse your test results and evaluate your system for conformance against the technical specification (and, hence, the client's brief).

> **You should:**
> - Demonstrate through your test results that **thorough** and **structured testing** has been carried out, including the simulation of some **unexpected events**.
> - **Link test results** to the **client brief**, to show a **fully supported** judgement of conformity of the system to the brief.
> - Produce an **audiovisual recording** to make a **judgement on the outcome** from many tests and the system in operation.
> - Complete a **log book record** for each **session**, with actions for the next session (see pages 89–100).

Test your system and analyse the results

> As you **test your system** and **analyse the results**, you need to make **records**. Read through the examples that follow on pages 122–123. Use the revision activities and guidance on pages 124–125 to apply them to your own project for **system testing** and **result analysis** for Revision activity 5.

Date: 26 June 2017 11:45	
General comments:	
I wrote the program to control the light intensity and adjusted the 3 LEDs so they gave the required outputs.	
Issues encountered and solutions, with justification:	
I had to adjust my program several times to get it working as needed. I had planned to use 3 different colour LEDs but they had very different levels of brightness. I changed my plan and decided to use white LEDs with colour filters over them. This was a quick solution and allowed me to move on.	
Action list for the next session:	
I will combine parts of program I have developed already to turn the LEDs on for the correct time and at the correct intensity.	

The circuit diagram below uses appropriate industry standards to accurately communicate technical information and operational requirements. The image is a test circuit built on a breadboard and shows the development processes have been carried out in a structured manner. **The flowchart is not annotated sufficiently for a third party to follow its structure**. Consider how you would improve the annotation.

This is part of the flowchart that will read the rotary switch and set the LED intensity:

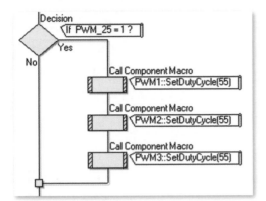

PWM flowchart

This is a picture of the circuit assembled on the microcontroller:

Rotary physical circuit

This is the circuit diagram for connection of the switch that will select the LED intensity:

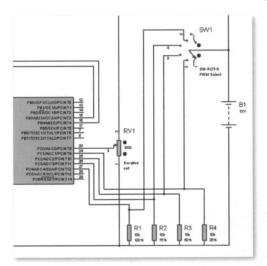

Rotary circuit diagram

Using Pulse Width Modulation (PWM)

Guided

For use with the next activity, research how to use **Pulse Width Modulation** (PWM) on a **microcontroller platform** of your choice and complete the points below. Then read the examples.

1 Explain what PWM is used for.

....

....

2 Describe how PWM works.

....

....

3 Explain the meaning of 'duty cycle'.

....

....

Guided example

Using a free light meter app for my mobile phone, I recorded the values of light intensity from the LEDs. I then made a guess of what values to use for the PWM duty cycle in my flowchart. I adjusted the duty cycle until a got the values I wanted.

This picture shows how I measured the light levels from the LEDs.

Test equipment set up using the Arduino™ Uno

You should **explain** in your **annotations** what the **images show**. You should also explain **why** you undertook the task and **what** you hoped to achieve.

When using tables, such as the ones that follow that show the test results, you need to **explain** what the **column headings** mean. You can find information on these headings on page 126.

Light level readings from first test:

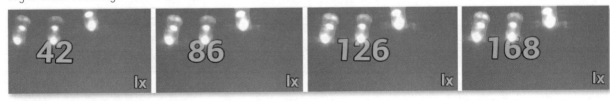

179 lx   108 lx   75 lx   40 lx

First measured readings

Initial

Duty	Actual lux	Target %	Actual %	Error	Target lux	Lux per duty	Calc duty
255	179	100	100	0.0	179.0	0.702	
150	108	75	60.3	–14.7	134.23	0.720	186
100	75	50	41.9	–8.1	89.5	0.750	119
50	40	25	22.3	–2.7	44.8	0.800	56

First test results.

Light level readings from second test:

Second

Duty	Actual lux	Target %	Actual %	Error	Target lux	Lux per duty	Calc duty
255	168	100	100	0.0	168.0	0.659	
186	112	75	66.7	–8.3	134.3	0.602	223
119	82	50	48.8	–1.2	89.5	0.689	130
56	40	25	23.8	–1.2	44.8	0.714	63

Second test results.

Light level readings from third test:

Third

Duty	Actual lux	Target %	Actual %	Error	Target lux	Lux per duty	Calc duty
255	158	100	100	0.0	158.0	0.620	
223	140	75	88.6	13.6	118.5	0.628	189
130	84	50	53.2	3.2	79.0	0.646	122
63	45	25	28.5	3.5	39.5	0.714	55

Third test results.

Light level readings from final test.

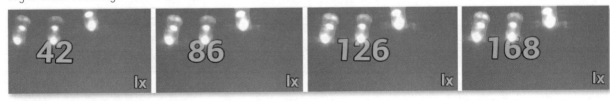

42 lx   86 lx   126 lx   168 lx

Final measured readings

Final

Duty	Actual lux	Target %	Actual %	Error	Target lux	Lux per duty	Calc duty
255	168	100	100	0.0	168.0	0.659	
189	126	75	75.0	0.0	126.0	0.667	
122	86	50	51.2	1.2	84.0	0.705	119
55	42	25	25.0	0.0	42.0	0.764	

Final test results.

Develop a system using Pulse Width Modulation (PWM)

 Use a **development platform** (i.e. **microcontroller** and associated **software**) of your choice to develop a system that uses **PWM** to control the brightness of an LED.

In order to complete this you should:

1. Write a functioning **program** that makes uses of **PWM** (see the example on page 127).

2. Use the program to **control** the **light intensity** of at least a single LED.

3. Establish a **method** of measuring the light intensity of the LED.

4. Use **PWM** to **adjust** the intensity to 25%, 50% and 75% of the maximum intensity.

5. Produce a **spreadsheet of the results**, and **calculate** the expected duty cycle required to obtain the specified levels. Print out your spreadsheet and paste it below as a record.

The spreadsheet should have the following columns:

a) Duty – This is the value you specified in the PWM command.

b) Actual lux – The measured light intensity.

c) Target % – What percentage of the maximum value you are trying to achieve.

d) Actual % – What percentage of the maximum light intensity is actually produced by the specified duty cycle.

e) Error – The difference between the actual percentage and target percentage.

f) Target lux – Calculations of what 25%, 50% and 75% of the light intensity should be.

g) Lux per duty – Calculates the lux produced per duty cycle.

h) Calc duty – Calculate what the duty cycle should be to achieve the desired output level.

My spreadsheet of results:

This is an **example** of a simplified **program** that will set the brightness of a single LED to one of two different values. It could be **extended** to meet the requirements of the **client brief**. It can be used as a **guide** for you to produce **your own version of the program** in a programming language of your choice.

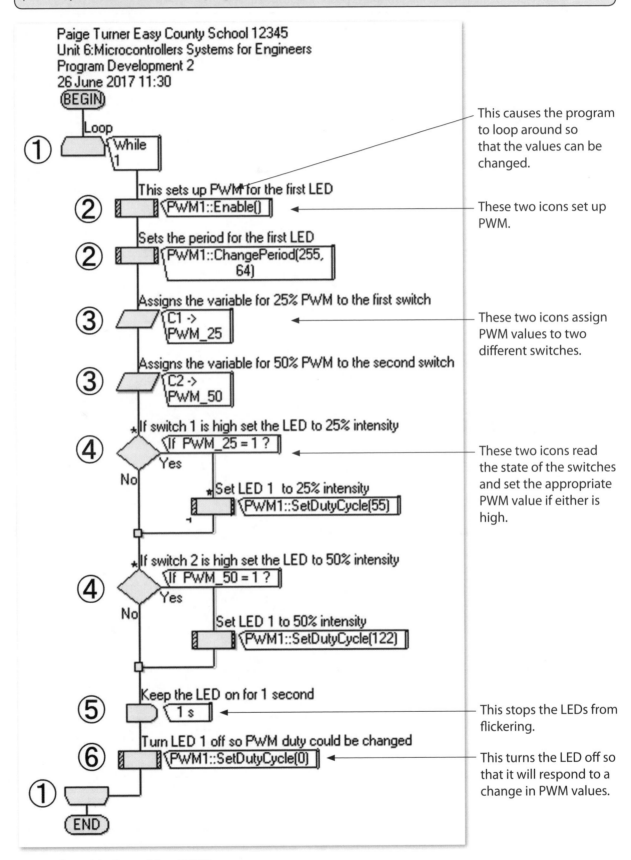

Paige Turner Easy County School 12345
Unit 6:Microcontrollers Systems for Engineers
Program Development 2
26 June 2017 11:30
BEGIN

① Loop While 1 — This causes the program to loop around so that the values can be changed.

② This sets up PWM for the first LED — PWM1::Enable() — These two icons set up PWM.

② Sets the period for the first LED — PWM1::ChangePeriod(255, 64)

③ Assigns the variable for 25% PWM to the first switch — C1 -> PWM_25 — These two icons assign PWM values to two different switches.

③ Assigns the variable for 50% PWM to the second switch — C2 -> PWM_50

④ If switch 1 is high set the LED to 25% intensity — If PWM_25 = 1 ? — Yes / No — These two icons read the state of the switches and set the appropriate PWM value if either is high.

Set LED 1 to 25% intensity — PWM1::SetDutyCycle(55)

④ If switch 2 is high set the LED to 50% intensity — If PWM_50 = 1 ? — Yes / No

Set LED 1 to 50% intensity — PWM1::SetDutyCycle(122)

⑤ Keep the LED on for 1 second — 1 s — This stops the LEDs from flickering.

⑥ Turn LED 1 off so PWM duty could be changed — PWM1::SetDutyCycle(0) — This turns the LED off so that it will respond to a change in PWM values.

① END

Example guide for writing PWM program

Program in your chosen programming language

Produce your **own version of the program** in a programming language of your choice, using the program on page 127 to guide you. Print it out and paste it below as a record.

Guided example

Date: 27 June 2017 11:25	
General comments:	
As all my programs were working, I tested the accuracy of the times that the LEDs stay on.	
Issues encountered and solutions, with justification:	
Instead of being able to set the time to 60 s, the maximum was 59. This is probably a hardware problem, but could be resolved by adjusting the software. When I finished these tests, I made the adjustment.	
Action list for the next session:	
Organise and produce a video recording of system in operation.	

Guided example

The guided example demonstrates that a **structured approach** to **testing** has been carried out. A better response would also include simulation of **unexpected events** (e.g. what would happen to the system if a LED failed), and a check of **conformity** with the **client brief** (see page 130 for an example of a conformity check response after the measure of the accuracy of timing 1 second). Make sure that you refer to the criteria for the task to ensure you meet all the requirements.

How I tested the times and PWM:

I connected one of the LEDs to an oscilloscope. To begin with I set the times to the minimum (1 s), maximum (59 s) and midway (30 s). I thought if the times were accurate at these points they would probably be accurate across the whole range. Then I set times so that I could look at the PWM signal.

This shows an example of one of the different settings used.

Settings for tests

An example of the result from testing how long the LEDs stay on for: this is the result for the 1 second test.

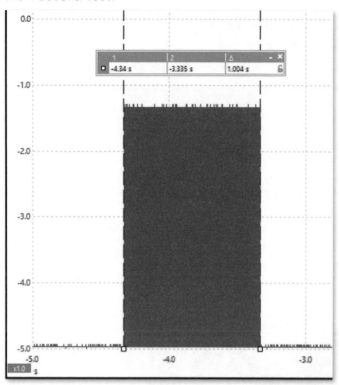

Measure of accuracy for 1 second

<u>Judgement of conformity</u>

All of the times were within a couple of milliseconds of what they should be.

This image shows the results from the 25% PWM setting. It shows the time is 322.2 µs. This should give the same levels of light intensity as determined in my previous tests. The results from 50% and 75% showed a similar degree of accuracy. Everything seems to be working as I want, so I think the system is now ready for me to start producing the final video.

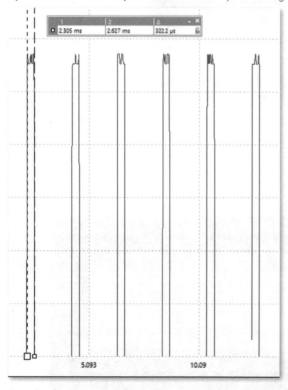

Results from testing PWM: measurement of PWM for 25% intensity

Record the outcomes of testing

You next need to produce an **audiovisual recording** of the **outcomes** of **testing**. It is important you are **familiar** with using a **range of equipment** to **test** performance and **record** results. You must develop the skills and knowledge needed prior to undertaking your actual assessment. Make sure you know the file types allowed for saving the recordings. Check with your tutor or look at the most up-to-date Sample Assessment Material on the BTEC National Engineering page of the Pearson website.

Using a system that you have developed previously, **test** and **produce** an **audiovisual recording** of its **performance**. Your tests should be **objective** and you should be able to **evidence the outcomes** of your tests.

Based on the **system shown above**, suggest possible tests for **unexpected events** that could occur. This results in an improved answer.

....
....
....
....
....

> Guided

Link the **results** from the above back to the **client brief** for a strong answer.

....
....
....
....
....
....
....
....
....
....

Links To revise test plans, see page 238 of the Revision Guide, and to revise analysis of test data and recording evidence, see page 239 of the Revision Guide.

Revision activity 6: System in operation

Make an evidential audiovisual recording that demonstrates your solution in operation with commentary that explains the operation of the system, and how its behaviour is linked with your chosen hardware and program.

The audiovisual recording of the system in operation should also provide visual evidence of the outcome from the testing you carried out.

> **You should:**
> - Produce an **audiovisual recording** that shows a **functioning system** that meets the **client brief**, while providing enhanced user experience and some handling of unexpected events.
> - Provide an audiovisual commentary that shows thorough understanding of **how the system operates** and the **relationship** between hardware and the program throughout.
> - Use **accurate** technical terminology.
> - Complete a **log book** record for each **session**, with actions for the next session (see pages 89–100).

Make a record of your solution in operation

> As you make a **recording** of your **system in operation** along with a **commentary**, you need to make **records**.

 Guided

 Guided example

 Read through the **examples** that follow on pages 132–141. Use the revision activities and guidance to **apply them to your own project** for the **system in operation** for Revision activity 6.

Date: 28 June 2017 11:35	
General comments:	
In this session I planned the content and how I would produce my video.	
Issues encountered and solutions, with justification:	
My video can last only 3 minutes and it is hard to appreciate how long my commentary will last. I rehearsed this against a stopwatch to get the time correct. This helps make sure I will use my time effectively.	
Action list for the next session:	
Record the audiovisual commentary of my system in operation.	

Plan content for the final audiovisual recording

Guided > The table below shows part of the **planned content** for the **final audiovisual recording**. Add more content so that the final recording will show all key features of the system in operation. You should estimate the time required for each part and ensure that the total is in the required time limit. Check with your tutor or the Sample Assessment Material on the Pearson website for the time limit in your actual assessment.

Content for the final video

Section	Content	Time (s)
1	Introduction: state my name, learner registration number and centre number	
2	Overview of user inputs	
3	Link input hardware to program	
	Total	

Plan the content of each section

Having structured the audiovisual recording, you should **plan the content** of each **section**.

Plan for user input section of recording		
Visual content	Audio content	Time
Point to time selection	"This is where the user sets how long the LEDs stay on for"	3
Point to intensity selection	"This is where the user sets how bright the LEDs will be"	3
Adjust time selection	"This is how the user adjusts the time, the LCD shows the setting"	3
Adjust intensity selection	"This is how the user adjusts intensity of the LEDs, the LCD shows the setting"	3
This allows a 1 second break between each part.		

Choose one of the sections you have identified for your presentation and plan its content using a format similar to the example above.

Plan for user input section of recording		
Visual content	Audio content	Time

Show the complete system operating

> **Guided example**

Your final audiovisual recording will need to show the **complete system operating**. It is important that it focuses on the **system**, and not the surrounding environment. You should try to exclude any information not related to the system or yourself.

Notice the focus in the image below which illustrates this point.

The final complete working system

Show the relationship between hardware and software

> Your final recording requires you to show an **understanding** of the **relationship between hardware and software**. One way of preparing for this would be to print out pages that **link** the **hardware** components of your system to the **software** that controls them. These can be referred to during the audiovisual recording. The images on pages 136–139 illustrate this approach.

Guided example 〉 Hardware and software to set time:

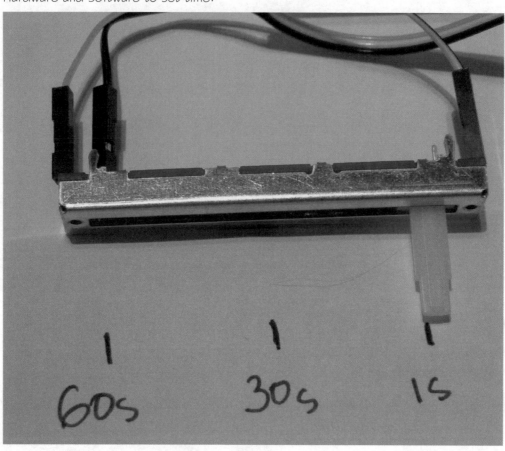

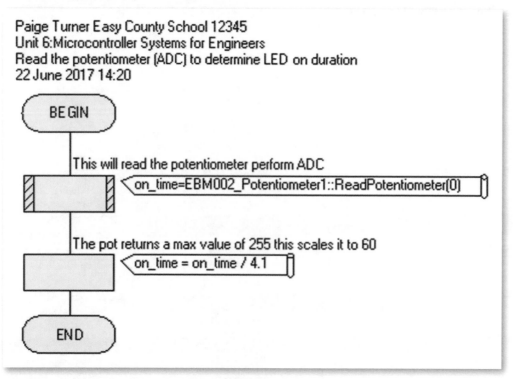

Hardware and software to set LED duration

Hardware and software to display information: On time is 1; Intensity 25:

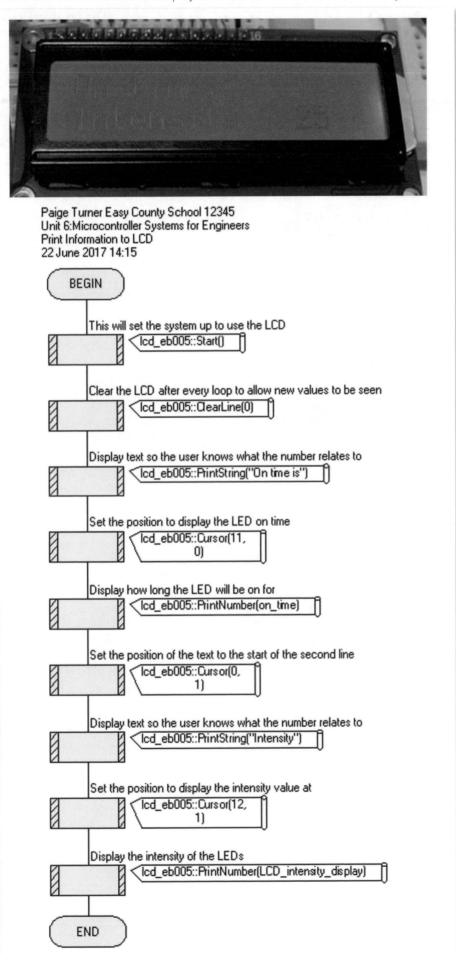

Paige Turner Easy County School 12345
Unit 6:Microcontroller Systems for Engineers
Print Information to LCD
22 June 2017 14:15

BEGIN

This will set the system up to use the LCD
lcd_eb005::Start()

Clear the LCD after every loop to allow new values to be seen
lcd_eb005::ClearLine(0)

Display text so the user knows what the number relates to
lcd_eb005::PrintString("On time is")

Set the position to display the LED on time
lcd_eb005::Cursor(11, 0)

Display how long the LED will be on for
lcd_eb005::PrintNumber(on_time)

Set the position of the text to the start of the second line
lcd_eb005::Cursor(0, 1)

Display text so the user knows what the number relates to
lcd_eb005::PrintString("Intensity")

Set the position to display the intensity value at
lcd_eb005::Cursor(12, 1)

Display the intensity of the LEDs
lcd_eb005::PrintNumber(LCD_intensity_display)

END

Hardware and software for LCD

Hardware and software to set intensity:

Paige Turner Easy County School 12345
Unit 6:Microcontroller Systems for Engineers
Read the switches to set PWM values
22 June 2017 14:15

BEGIN

Has the switch been set to 25%
If PWM_25 = 1 ?

No / Yes

Set this number as the duty cycle for PWM
intensity = 42

Display this % on the LCD
LCD_intensity_display = 25

Has the switch been set to 50%
If PWM_50 = 1 ?

No / Yes

Set this number as the duty cycle for PWM
intensity = 84

Display this % on the LCD
LCD_intensity_display = 50

Has the switch been set to 75%
If PWM_75 = 1 ?

No / Yes

Set this number as the duty cycle for PWM
intensity = 120

Display this % on the LCD
LCD_intensity_display = 75

Has the switch been set to 100%
If PWM_100 = 1 ?

No / Yes

Set this number as the duty cycle for PWM
intensity = 255

Display this % on the LCD
LCD_intensity_display = 100

END

Hardware and software to set PWM

Hardware and software to display light:

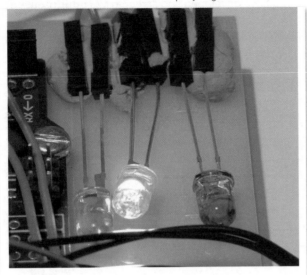

Paige Turner Easy County School 12345
Unit 6:Microcontroller Systems for Engineers
Output to LED at correct intensity
22 June 2017 14:10

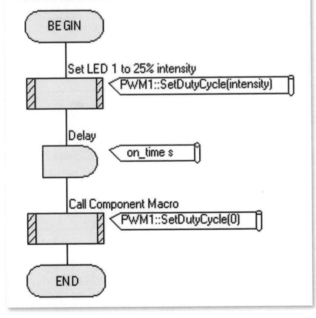

Hardware and software for PWM output

Combine **photographs of the hardware** you have developed and **screengrabs or printouts of the program** you have developed to produce images such as those in the previous examples. Use these to demonstrate your **understanding of the relationship** between **hardware and software** in your system. Paste your printout of descriptions and pictures as a record below.

System developments

Guided

> ✏️ Your **audiovisual recording** will need to take into account **enhanced user experience** and the handling of **unexpected events**. Note your suggestions below.

Suggestions for how the system that has been developed could be extended to provide an enhanced user experience.

....

....

....

....

....

....

Suggestions for how the system that has been developed could be extended to handle unexpected events.

....

....

....

....

....

....

The **development of the system** to fulfil the **client brief** is unlikely to proceed exactly as you plan at the outset. You should expect to make some **adjustments** and **changes** as the **system develops**. You will need to balance the time by completing each task to the standard you would hope for and ensuring that you **complete all activities**. You must remember this is a time-constrained task and there will probably be developments remaining that you would like to complete. This is likely to be so for most other people completing the task.

 Links To revise the audiovisual recording, look at pages 240–241 of the Revision Guide.

END OF TASK

Answers

Unit 1: Engineering Principles

Revision test 1

Section A: Applied mathematics (page 2)

1 $x = 0.81, y = 2.75$
2 1.5 s
3 109.1 m^2
4 0.485 s
5 15.43 m

Section B: Mechanical and electrical, electronic principles (page 7)

6 9270.5 N
7 (a) 1 103 625 N (1.1×10^6 N)
  (b) 1 839 375 N m (1.8×10^6 N m)
8 (a) Vertical support reaction at A = 35.33 kN, Horizontal support reaction at A = 10.39 kN
  (b) Vertical support reaction at B = 32.66 kN
9 0.213 m/s
10 (a) 40 Hz
  (b) 110 V
  (c) The RMS voltage (or current) is equal to an equivalent DC voltage (or current) that when connected across a resistor would produce the same heating effect. This is useful because RMS values can be used in Ohm's law and power calculations when dealing with resistive loads.
11 (a) 9200
  (b) 0.0435 A or 43.5 mA
  (c) Hysteresis losses, eddy currents
12 0.32 µF
13 (a) Labelled sketch of phasor diagram:

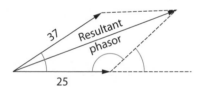

  (b) $v_R = 54 \sin(30\omega t + 0.635)$
14 0.82 V
15 0.00251 A or 2.51 mA
16 1.34×10^{-6} m^2
17 2682.4 kJ
18 203 kPa

Section C: Synoptic question (page 23)

19 (a) 16.9%
  (b) 91%
  (c) Full wave bridge rectifier circuit with smoothing capacitor

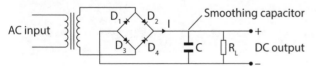

Revision test 2

Section A: Applied mathematics (page 25)

1 $l = 2T + 120$
2 $t = 2$ and $t = 4$
3 2.66 m^3
4 1.569
5 0.33 m^2

Section B: Mechanical and electrical, electronic principles (page 31)

6 (a) 13.86 kN
  (b) 17.85°
  (c) 0.398 m
7 16.83 MPa
8 2.22 m/s
9 11.54°
10 28 920 N
11 4.69 kΩ
12 214.50 mm
13 956 kg/m^3
14 (a) 1.6 Ω
  (b) 9.25 V
15 They must have low reluctance.
16 The commutator is part of the rotating armature in a DC motor. The commutator is split into several sections arranged in opposing pairs. Each pair is connected to a coil in the armature. Stationary carbon brushes keep contact with the commutator as it rotates. These feed current through the commutator to the coils in sequence as they rotate.
17 1 The number of turns in the coil.
  2 The speed of rotation.
  3 The strength of the magnetic field.
  4 The orientation of the coil to the direction of the magnetic field.
18 5.4 A
19 (a) 427.9 Ω
  (b) 0.538 A or 538 mA

Section C: Synoptic question (page 45)

20 (a) 1.557 MJ
  (b) 8.7%

Unit 3: Engineering Product Design and Manufacture

Researching and making notes

Planning your time (page 54)
Individual responses.
Identifying areas for research (page 54)
Individual responses.
Notes on the existing product design might include:
The jointing kit consists of two low carbon steel plates that are used to reinforce a butt joint between two wooden beams. The plates are fixed either side of the joint using six large bolts, nuts and washers. Before fitting you need to drill six holes, three each in the ends of beams to be jointed. The plate can be used as a template to position the holes. The plates are galvanised to prevent corrosion. The plates are manufactured in batches of 2000 by laser cutting. The plates and fixings must be suitable for use in the temperature range −40° to +50 °C.
The jointing kit will contain:
- 2 off 300 × 95 × 5 mm galvanised low carbon steel joining plates
- 6 off M12 medium carbon steel bolts and associated nuts, plain washers and spring washers.

Notes on areas of possible research might include:

- How the product could be improved.
- Optimising the existing design.
- Function of product; e.g. alternative products that do a similar job, ways of making joints in wooden beams.
- Manufacturing methods and processes; e.g. cutting out components from sheet materials, MIG/TIG welding.
- Costs; e.g. material costs, cost of equipment, cost of labour time required during manufacture or fitting the product.
- Health and safety issues; e.g. preventing components falling from height during installation, safe installation, eliminating sharp edges/corners, trapping hazards during installation.
- Sustainability and environmental issues; e.g. waste material or pollution caused during manufacture or use, efficient use of materials, use of recycled materials.
- Relevant numerical data; e.g. ways of analysing data, measures of central tendency, interpreting data.
- Aesthetics, ergonomics, anthropometrics; e.g. physical limits on product size and weight that can be safely handled at height by one person, shape.
- Joints/fixings; e.g. nuts, bolts, washers, coach screws, nails, woodscrews, adhesives.
- Applied finish; e.g. zinc plating, chrome plating, painting, powder coat, galvanising.
- Advantages/disadvantages of the design: cost, ease of installation, sustainability, safety.

Focus for further product research (pages 55–58)

Individual responses. Research and notes must be relevant to the task brief and address the categories on pages 55–56:

- Choice from a wide variety of wooden beam joining products available on the internet
- Image of possible solution and function of product
- Possible manufacturing methods and alternative methods
- Approximate cost
- Available materials and alternative materials and properties
- Health and safety issues – processes and products, designing out risk
- Sustainability and environmental factors
- Relevant numerical data
- Aesthetics, ergonomics, anthropometrics
- Applied finish
- Advantages
- Disadvantages
- Sustainability
- Other relevant information.

Researching materials, processes, fixtures, fittings and relevant numerical data (pages 59–60)

Individual responses. Research, notes and sketches must be relevant to the set task brief. Your research notes for the example tasks might include some of the following.

Roof beam materials and beam dimensions

The most popular wood species for roof joists and rafters is kiln dried spruce. The source of the wood should be FSC certified so that you know it is being grown and harvested sustainably. In the UK construction industry, beam dimensions are strictly specified according to the load they are expected to support. It also depends upon the grade of timber used. For example, 47×97 mm C16 Grade timber (with density 0.75 kN/m^2) can span a maximum of 1.71 m, whereas a 47×220 mm piece can span 4.6 m, both only for flat roofs. Common standard sizes are still often referred to in the old imperial sizes like 2×4" or 2×8". Common metric sizes range from 47×100 mm, 47×125 mm up to 47×250 mm. Spruce timbers are stable once dried, are renewable and actually capture and store carbon.

Nail plates

Nail plates are the most popular joist joining method in the UK. Most are manufactured in galvanised 1 mm steel plate. They are most commonly used for connecting beams and for repairs. Nail plates are punched to create small multiple prongs on one side. These are hammered into the surfaces of adjacent pieces of timber and act like nails to hold them together.

Low carbon steel, aluminium alloy duralumin, austenitic stainless steel alloy

The materials used for the joining plates might have to operate in temperatures as low as –40 °C, depending where in the world the building is situated.

My research has shown that low carbon steel can lose some of its impact strength at extremely low temperatures, making it brittle. Steel operating below the ductile to brittle transition temperature is far more likely to fail under shock loading.

However, duralumin and austenitic stainless steel alloys retain their impact strength and resistance to shock loading even at very low temperatures.

Austenitic stainless steel: 300 Series austenitic stainless steel has austenite as its primary phase (face-centered cubic crystal). These are alloys containing chromium and nickel, and sometimes molybdenum and nitrogen, structured around the Type 302 composition of iron, 18% chromium and 8% nickel. The advantage of using this material is its very high tensile strength compared to carbon steel.

Aluminium is the most abundant metallic element on Earth and is the most widely used non-ferrous metal in engineering. It is ductile and highly malleable, which makes it suitable for drawing into wires or rolling into thin foil. However, pure aluminium does not have the required tensile strength for most structural applications unless it is alloyed with other metals. Duralumin is an alloy of aluminium that is strong, hard and lightweight. It is mainly used in the aircraft industry. After heat treatment and ageing, Duralumin is comparable to steel in terms of strength but with considerably less weight. Aluminium loses its corrosion resistance when alloyed to Duralumin and therefore Duralumin has a special laminated coating applied, called Alclad. This thin surface layer of pure aluminium covers the strong Duralumin core.

Fasteners such as socket cap screws, structural bolts, washers and coach bolts

Socket cap screws are commonly used in machine parts. The socket head enables driving where there is not sufficient space for spanners or sockets. They have a small cylindrical head with tall vertical sides. Allen (hexagon socket) drive is a six-sided recess for use with an Allen key (hexagon key).

Structural bolts used in construction have high tensile strength and enlarged heads that, when used in conjunction with plain washers, help to spread load on the surface of beams.

A coach bolt (sometimes called carriage bolt or round head square neck bolt), is a form of bolt used to fasten metal to wood. It has a shallow mushroom head and square section on the shank immediately under the head. This makes it self-locking when inserted into a square hole in a fixing plate or bracket. It can also be installed using only a single tool, spanner or socket.

Plain washers are used to spread the load exerted by a nut or bolt head on the surface onto which they are tightened.

Spring washers are used to prevent the movement of nuts once they are tightened and so prevent vibration from causing them to loosen over time.

MIG and TIG welding processes

Gas metal arc **welding** (GMAW), sometimes referred to as metal inert gas (**MIG**) **welding**, is a **welding** process in which an electric arc forms between a continuously fed consumable wire electrode and the workpiece metal(s), which heats the workpiece metal(s), causing them to melt and join. The inert gas shield around the weld prevents oxidation and improves the quality and reliability of the weld. MIG welding is best suited to steel but by changing the consumable wire aluminium and stainless steel can also be welded, although this tends to be more difficult.

Gas tungsten arc **welding** (GTAW), also known as tungsten inert gas (**TIG**) **welding**, is an arc **welding** process that uses a non-consumable tungsten electrode to produce the weld.

A consumable metal rod is fed into the weld by hand to form the joint. TIG welding is commonly used for otherwise difficult metals to weld, such as aluminium alloy.

Reviewing further information

Outcomes of simulations and testing on existing plates (page 64)
1. Average expected life cycle of the product: 35.6 years
2. Drilled holes are 13.5 mm diameter.
3. Main client motivation for the redesign is because the plates are fracturing in service in certain environments.

Responding to activities

Revision activity 1: Planning and design changes made during the development process (pages 65–67)

Project time plan: Individual responses. The project time plan must show that there is a logical and iterative approach to the design process.

Changes and action points for each session: Individual responses.
- The design development must link to research and the requirements of the brief.
- All design change developments should be justified.
- Action points for the next session should be identified, that are clear and logical, prioritised and Specific, Measurable, Achievable, Realistic, and Time-based (SMART).

Revision activity 2: Interpret the brief into operational requirements (pages 68–69)

Individual responses.
- The product requirements should be cohesive and comprehensive.
- The opportunities and constraints should be feasible and meet the brief, enhancing product performance.
- The calculation and interpretation of numerical data should be accurate.
- The health and safety, regulatory and sustainability factors should be relevant to the given context.

Interpreting the data in Table 1 (page 69)
- The average life cycle in years for plates used in the United Kingdom is 37.25 years.
- The average life cycle in years for plates used in Norway is 26.4 years.
- The country where beam plates have the shortest average life cycle is Alaska.
- The country where beam plates have the longest average life cycle is Brazil.
- The temperature range the plates are subjected to, is least in U.K.
- The temperature range the plates are subjected to, is the biggest in Alaska.
- The plates are subjected to the lowest average humidity in Norway and Alaska.
- The data would seem to suggest that the likely cause of premature failure of the beam plates is the colder temperatures and/or the temperature range the plates are subjected to in service.
- This may be because mild steel tends to become more brittle at lower temperatures.
- Additional information from the numerical data could include whether humidity is relevant to the results, or the fact that the plates do not meet the designed life cycle of 50 years, in any of the locations.

Revision activity 3: Produce a range of initial design ideas based on the client brief (pages 70–73)

Individual responses.
- Your ideas should be sketched clearly and solve the problem stated in the client brief.
- You should annotate your sketches with technical and engineering terms.
- All your ideas should be realistic and suitable to address the task you have been set.

Revision activity 4: Develop a modified product proposal with relevant design documentation (pages 74–78)

Individual responses.
All development work should show that you have met the checklist of 10 important categories on page 74, including the following:
- The solution should be developed to be the best possible response to the brief that you can make it. You must show that you have produced a different design than the original, and state clearly why you have made the alterations you have made. The design proposal should show your engineering research and knowledge based on a thorough understanding of existing alternative products.
- Material(s) selection should be suitable to solve the problem as set in the brief, and you should give reasons why you have chosen these materials. You should also show that you have properly researched all the possible options available. Selection of manufacturing process should be suitable for making the design you have chosen and backed up with thorough research of alternative methods. You should clearly state why you have made the decision you have arrived at.
- The design proposal should consider sustainability at all stages of the product life cycle. Ideas should clearly mention the safety of the design and how you have designed out risks.
- Your documentation should include plans, with all dimensions and manufacturing methods clearly listed. A good engineer should be able to make your design just from your paperwork. You should use correct technical language all the way through your assessed task booklet.

Revision activity 5: Evaluate the design proposal (pages 79–81)

Individual responses.
Your evaluation should show that you have considered the stronger and weaker points of your design proposal. You should be clear about the positives and negatives, always giving evidence to back up your statements.

You should present fair and balanced statements to explain why your design is a good solution to the problem set in the original brief. You should also clearly explain how your design could be further improved as a result of any new technologies becoming available in the future. If there are any problems with your design, you should show that you recognise what they are, and suggest practical ways improvement could be made.

Unit 6: Microcontroller Systems for Engineers

Revision task

Understand the scenario, client brief and task (pages 86–88)

Individual responses. Underlining of the client brief may include the example below.

The scientists want to determine the effects of several factors. They want to observe if the <u>colour of the light</u> is an important factor in the <u>growth of plants</u>. They have decided that they want to compare how seedlings react to <u>red</u>, <u>green</u> and <u>white</u> light. Another factor they believe may influence the seedlings' growth is the <u>intensity of the light</u>.

In order to test the response of the seeds to the colour of light the experiment requires that a seed will be exposed to a <u>repeating on/off sequence</u> of the different coloured lights. Each light should be on for a <u>user-selectable duration</u> of between <u>1 and 60 seconds</u>. It is expected that the experiment may take up to <u>10 days</u> to show measurable results. During this period the seeds will be <u>excluded from ambient light sources</u>.

In order to test the response of the seeds to the <u>intensity of light</u> each light should be able to have its <u>output varied in increments of 25%</u>. Each light should <u>emit the same intensity</u> of light during a single experiment.

Each of the light sources should be <u>50 mm above the seed, 15 mm from the centre of the seed. Each light source should be separated by 120°</u>. The light sources should <u>not affect the environmental temperature</u> of the experiments.

You have been allocated a budget of <u>£10.00</u>, excluding the microcontroller hardware itself.

Key questions on the client brief (page 86)

Example answer:

1 What inputs are required?
   A method of setting the duration the lights stay on for.
   A method of controlling the intensity of the lights.

2 What function are the inputs required to perform?
   To control how long the lights will be on for – it will need to be able to select values between 1 and 60 seconds.
   The lights will need to have their intensity varied between 0%, 25%, 50%, 75% and 100%.

3 What outputs are required?
   There will need to be three separate different coloured light sources.

4 What functions are the outputs required to perform?
   There will need to be red, green and white lights of the same intensity.

5 What aspects of the client brief relate to user experience?
   The user will want to know how long they have set the lights to stay on.
   The user will want to know the intensity of the lights they have set.

6 What constraints are imposed?
   £10 budget for cost of components.
   The system must not affect the environmental temperature of the experiment.
   The system must be able to run for 10 days.

Revision activity 1: Task planning and system design changes

Project time plan (page 87)

Individual responses.

Initial task plan (page 88)

1 Read through the assessment task to understand what is required.
2 Note down the sessions needed to complete the work.
3 Analyse the scenario and client brief.
4 Produce the initial project time plan.
5 At the start of each session review plan from previous one.
6 At the end of each session plan what to complete for the next one.
7 During every session review progress against timetable.
8 Choose the hardware to use.
9 Design interface between hardware and microcontroller.
10 Plan what the program should do.
11 Plan how to test system.
12 Produce initial programs for each subsystem of the task and simulate using software.
13 Produce first version of complete program and simulate using software.
14 Assemble hardware for a single subsystem of the task.
15 Program microcontroller and test a single hardware/program subsystem; e.g. focus on setting light on duration.
16 Resolve any hardware, microcontroller or program problems with the single subsystem and move to the next subsystem, correcting problems as required before moving on.
17 Test and record operation of complete system.
18 Make audiovisual recording of operation of system.

Task session log book tables (pages 89–100)

Individual logbook tables for each task session covering Activities 1–6.

Revision activity 2: Analysis of the brief

Identify the key requirements your solution must meet (page 101)

Key requirements are:

1 Red, green and white lights
2 User-selectable duration of between 1 and 60 seconds
3 £10.00 budget for components
4 Vary intensity of the light
5 Repeating on/off sequence
6 10 days' duration of experiment
7 User-selectable light intensity output varied in increments of 25%
8 Emit the same intensity of light
9 Not affect the environmental temperature.

Interpret the brief into operational requirements (pages 102–103)

Individual responses reflecting development platform, components and operational requirements for individual solutions.

Produce a test plan (page 104)

Individual responses reflecting individual solutions.

Record the outcome of your results (pages 105–106)

Individual responses reflecting individual solutions.

Revision activity 3: System design

Develop your solution and connection table (page 108)

Individual responses specific to the microcontroller used at each centre.

Below is an example based on the Arduino™ Uno.

Pin	Port	Function 1	Function 2	Connected to?
1	D3	Digital 3	Interrupt 1	LCD D6
2	D4	Digital 4		LCD D5
9	D5	Digital 5	PWM	LCD D4
10	D6	Digital 6	PWM	LED 1
11	D7	Digital 7		
12	B0	Digital 8		
13	B1	Digital 9	PWM	LED 2
14	B2	Digital 10	PWM	LED 3
15	B3	Digital 11	PWM	LCD E
16	B4	Digital 12		LCD RS
17	B5	Digital 13		
23	C0	Analogue 0		Set time
24	C1	Analogue 1		25% Intensity
25	C2	Analogue 2		50% Intensity
26	C3	Analogue 3		75% Intensity
27	C4	Analogue 4		100% Intensity
28	C5	Analogue 5		
29	C6	Reset		
30	D0	Digital 0		
31	D1	Digital 1		
32	D2	Digital 2	Interrupt 0	LCD D7

Select specific components (page 109)
Individual responses reflecting individual solutions.

Create a schematic diagram (page 111)
Individual responses reflecting individual solutions.

Create an outline plan for the program structure (page 112)
Individual responses reflecting individual solutions.

Revision activity 4: System assembly and programming
Produce your own version of the system (page 119) and
A solution in your chosen programming language (pages 119, 121)
Individual responses reflecting individual choices of development platform and programming language.

Revision activity 5: System testing and result analysis
Using Pulse Width Modulation (PWM) (page 124)
1 PWM is used to simulate an analogue output. It can be used to control how bright LEDs are, or how fast a motor turns. There are other uses.
2 PWM works by turning a digital output on and off rapidly.
3 Duty cycle is the relationship between the time the out signal is on and off.

Develop a system using Pulse Width Modulation (PWM) (page 126)
Individual responses depending on the platform of choice. The example below is based on the Arduino™ Uno.

Program in your chosen programming language (page 128)
Individual responses reflecting individual choices of program.

Record the outcomes of testing (page 131)
Individual responses.
Example for unexpected events might include:
As the system will be working for several days it is possible that hardware will fail. Once the system is running, components such as the LCD screen, potentiometers to set time and screen brightness or the switch to control the intensity could be removed. These should not affect the LED on/off sequence. In this way, even though the user interface would be compromised, the actual experiment would still be able to continue.
Example for linking the results back to the client brief might include:
The client brief requires that the lights should be able to be switched between 25%, 50%, 75% and 100% intensity. The results from this and previous tests confirm that the user will be able to accurately set the intensity of the LEDs to these values. The LCD screen will provide the user with feedback.
The client brief requires that the lights should be able to be turned on from 1 to 30 seconds. The results from this test confirm that the user will be able to accurately set the time to integer values between 1 and 60 (with some minor adjustments to get the full 60 seconds). The LCD screen will provide the user with feedback. The client brief does not indicate if fractions of a second are required but I have decided that this would not be needed. If the client did want to be able to set the time to fractions of a second it would require a calculation to be adjusted. This would allow 0.25 second intervals to be achieved.

Revision activity 6: System in operation
Plan content for the final audiovisual recording (page 133)
Individual responses. The stages might include those presented below; for example:

Section	Content	Time (s)
1	Introduction to me	
2	Overview of user inputs	
3	Link input hardware to program	
4	Overview of LCD display	
5	Link LCD control to program	
6	Demonstrate setting LED duration	
7	Link setting LED to program	
8	Demonstrate setting LED intensity	
9	Link LED intensity to program	
10	Review what information on LCD means	
11	Observe LEDs repeating their pattern at 1 second duration	
12	Observe LEDs at four different levels of intensity for 1 second	
13	Change duration LEDs stay on and observe	
	Total	

Plan the content of each section (page 134)
Individual responses reflecting individual solutions.

	A	B	C	D	E	F	G	H
1	Initial							
2	Duty	Actual lux	Target %	Actual %	Error	Target lux	Lux per duty	Calc duty
3	255	179	100	=B3/179*100	=D3-100	=B3	=B3/A3	
4	150	108	75	=B4/179*100	=D4-75	=$B$3*0.75	=B4/A4	=F4/G4
5	100	75	50	=B5/179*100	=D5-50	=$B$3*0.5	=B5/A5	=F5/G5
6	50	40	25	=B6/179*100	=D6-25	=$B$3*0.25	=B6/A6	=F6/G6

Show the relationship between the hardware and the software (page 140)

Individual responses reflecting individual solutions.

System developments (page 141)

Individual responses for enhanced user experience, depending on solutions. Example answer:

The system that has been developed could be extended to show the user how long it has been running for; i.e. total number of hours or day.

Setting the duration requires fine tuning the position of the potentiometer. A digital method, such as a keypad, would make this better.

Individual responses for handling unexpected events, depending on solutions. Example answer:

The system that has been developed could be extended to handle unexpected events. For example, the system is designed to run from a power supply. This could fail. If a battery backup was available this would prevent the experiment from becoming invalid.

There is no redundancy in the hardware. If any part fails the system would be compromised. If each component were duplicated any failures would not be a problem.

Notes